International Markets and Global Firms

SAGE STUDIES IN NEO-CORPORATISM

Series Editors: Philippe C. Schmitter and Alan Cawson

International Markets and Global Firms

A Comparative Study of Organized Business in the Chemical Industry

edited by
Alberto Martinelli

SAGE Publications
London • Newbury Park • New Delhi

First published 1991

SAGE Publications Ltd
6 Bonhill Street
London EC2A 4PU

SAGE Publications Inc
2455 Teller Road
Newbury Park, California 91320

SAGE Publications India Pvt Ltd
32, M Block Market
Greater Kailash - I
New Delhi 110 048

British Library Cataloguing in Publication Data

International markets and global firms: a comparative study of organized business in the chemical industry.
- (Sage studies in neo-corporatism)
1. Chemicals. Production
I. Martinelli, Alberto
388.4766

ISBN 0-8039-8436-7

Library of Congress catalog card number 90-53674

Typeset by Mayhew Typesetting, Bristol, England
Printed in Great Britain by Billing and Sons Ltd, Worcester

Contents

Preface

This book is the result of several years' research on the relationship between business and politics. During these years many intellectual debts have been accumulated.

This book is, first of all, one of the planned volumes of the nine-country comparative international project on the organization of business interests, coordinated by Philippe Schmitter and Wolfgang Streeck at the European University Institute in Florence. This ten-year project was the first major international comparative study on business associations; it was based on a rather elaborate theoretical framework and implied a complex organizational network composed of nine national research teams. Philippe and Wolfgang were outstanding coordinators, both in intellectual and in organizational terms. The quality of the members of the research team was equally outstanding.

The first idea for the project was discussed by Philippe, Wolfgang and myself in the summer of 1978 at the International Sociological Association Congress in Uppsala. A first theoretical framework was worked out by the three of us in the autumn of 1979 and was discussed at the 1979 Berlin Conference. Then, in 1980, Philippe and Wolfgang had the opportunity to spend several months together in Berlin, at the Internationales Institut für Management und Verwaltung, where they produced a well-developed research design which laid the groundwork for the empirical research. Several meetings took place during the following years to discuss the first results and to refine the major concepts, theories and hypotheses. These meetings involved the members of the research team and a small number of invited scholars, including Leon Lindberg and Rogers Hollingworth of the University of Wisconsin–Madison and Victor Perez Diaz of the University of Madrid. Throughout the work Philippe and Wolfgang acted as 'democratic leaders', convinced of the soundness of their ideas, but at the same time able to and willing to recognize the views and insights of others. The results of this major endeavour include, in addition to this volume, the book on the organization of business interest associations by Schmitter and Streeck which is forthcoming, the already published volume on the food processing industry edited by Wyn Grant (*Business Interests,*

Organizational Development and Private Interest Government, Berlin: de Gruyter, 1987), a volume on the construction industry and a volume on regionalism and business interest representation.

Although the book represents one of the 'sector studies' to emerge from the project, it has a very distinctive quality, which stems from the specific features of the chemical industry. The volume stresses such issues as the role of global firms as political actors, the impact of government industrial policies, the role of the European Community, etc.

The volume also draws on the research conducted at the University of Warwick in the Department of Politics, which was part of a comparative project on Government–Industry Relations in the Chemical Industry in the United Kingdom and West Germany, and within the EEC. This project was sponsored by the Economic and Social Research Council.

The ideas developed in the introduction, and in some of the chapters were discussed with members of my research group on the 'relations between the economy and society' at the University of Milan, Faculty of Political Science, and with scholars who participated in the Workshop on 'State, Market, and Social Regulation in Italy' in Bellagio in 1987. For the analysis of Montedison's strategy in Chapter 4, I am indebted to Claudio Dematté of Bocconi University.

The international coordination of the comparative project on the organization of business interests was funded by the Volkswagen Foundation. However, the funding of the country studies was provided by national institutions, such as the Ministry of Education in Italy, and the Nuffield Foundation in the United Kingdom.

ASAP, the business association of ENI, provided funds both for the Rome 1985 Workshop on Interest Representation in the Chemical Industry where the ground was laid for this volume, and for some of the empirical work for the Italian contributions to the volume. ASAP officials in Rome, and Federchimica officials in Milan, as well as business associations and trade union representatives in various countries and in the EEC, were also open and cooperative.

Last but not least, I owe a particular debt to Wyn Grant of the University of Warwick, who provided precious assistance throughout the entire gestation of the book.

ALBERTO MARTINELLI
Milan

Notes on the Contributors

Pamela Adams is Assistant Professor of Marketing and Vice Coordinator of the Master in International Economics and Management at the Graduate School of Business Administration, Bocconi University, Milan.

Antonio M. Chiesi is Assistant Professor of Sociology at Bocconi University, Milan. He has recently written (with A. Martinelli) 'The representation of business interests as a mechanism of social regulation', in P. Lange and M. Regini (eds), *State, Market and Social Regulation* (1989).

Wyn Grant is Chairman of the Department of Politics and International Studies at the University of Warwick, UK. His most recent books are *Government and Industry: A Comparative Analysis of the US, Canada and the UK* and *Pressure Groups, Politics and Democracy in Britain.*

Henry J. Jacek is Associate Professor in the Department of Political Science, McMaster University, Ontario, Canada. He is co-editor of *Regionalism, Business Interests and Public Policy* (1989).

Alberto Martinelli is Dean of the Faculty of Political Sciences and Professor of Political Science and Sociology at the University of Milan, Italy. He is chairman of the International Sociological Association Research Committee on Economy and Society. He is co-editor (with H. Makler and N.J. Smelser) of *The New International Economy* (1983) and (with N.J. Smelser) of *Economy and Society: Overviews in Economic Sociology* (1990).

Giuliano Mussati is Professor of Industrial Economics at Rome-Tor Vergata University and at Bocconi University in Milan. He is Co-Director of the 'Furio Cicogna' Research Center on Entrepreneurship, Bocconi University, and Economic Adviser of the Minister of Industry in the Italian Government.

William E. Paterson, formerly Professor of Politics, University of Warwick, is now Salvesen Professor of European Institutions and Director of the Europa Institute, University of Edinburgh. He is co-author with Wyn Grant and Colin Whitston of *Government and the Chemical Industry* (Oxford 1989).

Anna Soru is Senior Researcher at the 'Furio Cicogna' Research Center on Entrepreneurship, Bocconi University, Milan, after working for the Italian Chemical Manufacturers Association.

Colin Whitston is a Senior Research Fellow in the Industrial Relations Research Unit at the University of Warwick. He is co-editor (with Stephanie Tailby) of *Manufacturing Change: Industrial Relations and Restructuring* (1989).

1
Introduction

Alberto Martinelli

The general issues

The 1980s have been characterized by extensive technological innovation, important sectoral shifts in the economy and widespread social transformation. Some scholars argue that the United States and the United Kingdom, as well as Japan, have been in their unique ways more capable of dealing with the economic problems of these times than the other major advanced industrialized nations. The economic performance of the American economy in recent years – its achievements in reducing inflation, creating new jobs, modernizing industry and upgrading competitive capacity in key sectors – and the still more recent performance of the British economy – with high growth rates and sustained labour productivity increases – have impressed continental Europeans as much as the solid performance of Japan.

These achievements have been accompanied by problems such as trade deficits and public debt in the United States and high unemployment and social unrest in Britain. Still they have succeeded in raising the question of applying the 'Anglo-American' model to the problems of economic rationalization and international competitiveness in continental Western Europe. Doubts have been cast, in fact, about the ability of European industry to sustain high levels of employment and social services while remaining competitive with the United States and Japan.

The chemical industry, which is the specific object of the analyses in this volume, provides an excellent testing ground for such hypotheses.

The economic and industrial policies that characterize what we call the 'Anglo-American' model of the 1980s are different versions of the 'company-led' model of industrial adjustment in which the basic choices are made by individual firms without outside interference (Zysman, 1983). A company-led strategy has worked only in the United States, and even there it has faced problems in those parts of the country where obsolete industrial structures required the intervention of state governments.

Although the ultimate goal of the Thatcher government was to create the conditions in which company-led adjustment is possible, state intervention has been necessary in Britain because companies have lacked the political will and the managerial determination to see adjustment through on their own (Grant, 1987). Still in both cases there has been an emphasis on the free market and on the political use of the market to curtail the power of organized labour.

In continental Europe conditions have been somewhat different. While in the Anglo-Saxon countries there has been a mix of company-led industrial adjustment and state intervention, in continental Europe there has been a mix of state intervention and concertation, tripartite negotiations and company strategies accompanied by a greater level of institutional complexity.

Institutional complexity, in fact, has played an important role in continental Europe's response to the problems of governing an advanced capitalist economy and a post-industrial society within the framework of political democracy. This role has involved two basic sets of processes. The first has been an expansion of the role of the state through Keynesian demand-management policies, control of the business cycle, and welfare policies directed towards income distribution and the provision of social services. The second has involved the creation of a dense network of well-organized self-governing interest associations with multiple institutional arrangements for negotiation and intermediation between public and private interests. Institutional complexity and density were by no means the only possible responses to socioeconomic change, but they proved to be a viable solution for many European countries by guaranteeing high rates of growth and employment as well as high levels of political legitimation and social consensus.

Since the mid-1970s, however, persistent stagflation, increasing competition and rapid technological change have made elements of that 'postwar settlement' obsolete. The diminished performance of the European economies has reduced the resources available for welfare redistribution and for the enlargement of social citizenship. As a result policy-makers have shifted their attention from demand-management to supply-management policies that rely on direct control over the factors of production. Under these new conditions reliance on market mechanisms seemed to provide a more viable way to introduce technological change and catch up with international competitors. Institutional density, complex bargaining, and compromise concertation appeared more and more as a source of resistance to further innovation and modernization.

Yet such a radical change in policy presented many of these

countries with a tremendous challenge. They lacked some of the basic prerequisites of the American model of market adaptation and company-led adjustment. They also enjoyed less social mobility, less freedom of movement for capital and labour, and more consolidated expectations of income and job security – all protected by entrenched associations and entitlements. Any straightforward application of the American model would have entailed high social costs, such as the growth of an underclass in central cities, and widespread social conflict, which most European countries were not willing to sustain. The electoral systems of most European countries, moreover, would not have allowed the formation of a political coalition strong enough to impose a programme of massive reprivatization, market forces and political defeat of trade unions.

The central question, then, is has the 'European model' of state intervention and institutional density become obsolete, or can it be updated to provide the necessary structural change and competitive strength without endangering existing levels of social and political consensus? In other words, have the consolidated social and political assets of Western Europe, such as associational development, functional representation of organized interests and neo-corporatist bargaining become liabilities in the governance of the contemporary European economy?

The answer to this question has varied across Europe. The most extensive policy change has come in Britain. The Thatcher government attempted to shift the economy from a mixed to a market form by reducing state intervention, curtailing union power, exposing a free market ideology and showing a remarkable disposition to pay the social costs of such operations. The response in continental Europe has been more cautious. Two major orientations may be identified. France, West Germany, and Italy together with the other, smaller countries, introduced changes in the direction of a revitalization of the market. They did not, however, dismantle the complex institutional network that had been developed in earlier decades. Each country relied on its own strengths: West Germany relied on its coherent and centralized system and on the invisible and effective cooperation between the private and the public sector (Katzenstein, 1985); France made use of its competent and autonomous state bureaucracy which was able to guarantee a sense of continuity in the face of significant policy changes; Italy relied on entrepreneurial vitality and flexibility and on local systems of small firms. The process of European economic integration was also accelerated by the adoption of a set of measures aimed at creating a unified European market for goods, people, services and capital, and eliminating the physical, technical and fiscal barriers within Europe by the end of 1992.

In light of these policies one may argue that neither the thesis of a growing, autonomous role of the state in the economy which was developed by Shonfield in the mid-1960s (Shonfield, 1965), nor the renewed focus on the centrality of the role of the market in supply-side economics explains completely the complexity of the institutional responses of Western Europe to the problems of economic regulation and industrial adjustment. Other institutional factors, in fact, must be taken into account.

One of these factors is the role of business interest associations as private interest governments that perform tasks that would otherwise have to be undertaken by the state or by large firms. The central hypothesis of this volume, in fact, is that economic governance and industrial adjustment are the outcomes of a complex mix of institutional mechanisms of which organized business interests are a significant, if not predominant, part (Chiesi and Martinelli, 1989). In light of this general hypothesis, this volume aims to analyse the more specific topic of industrial change in the chemical industry. Focus will be placed on how an industrial market and a policy subsystem are governed by different political and economic actors, and how their interaction affects industrial restructuring. The analysis is based on the results of an international research project on the structure and activities of business interest associations in the chemical industry, and on the role played by self-governing associations in relation to other institutional actors such as individual firms, national governments, trade unions and supranational institutions. One of the major conclusions of the volume is that Europe has been successful, albeit at different degrees in the different countries, at coping with the crisis that hit the chemical industry in the 1970s, and at employing a balanced mix of coordinating institutions (including large firms, international markets, business associations and government agencies) to do so.

A conceptual framework for the study of business interest associations

Having outlined the general issues of interest in this volume, I will now summarize some of the basic concepts that have been applied in most of the chapters. These concepts are drawn from the literature on interest organizations and the relationship between business and the state and, more specifically, from the framework developed by Philippe Schmitter, Wolfgang Streeck and the other members of the international comparative research project on business interest associations (Schmitter and Streeck, 1981). Some significant modifications to this framework, however, will be proposed. I will then clarify why I have chosen the chemical industry as an appropriate testing ground for our research.

The international comparative project on organized business interests grew out of an observation that the efforts of business to advance and defend its interests had attracted surprisingly little attention from scholars, especially when compared with the literature on trade unions. The main themes of the research may be summarized in three general questions, all of which relate to the organizational dimension of business interest politics:

1 Why should the owners of capital, possessing as they do discretionary power to invest, need collective interest representation, and what are the societal conditions that determine the significance of the use of associations by business in the pursuit of its interests?
2 What structural conditions are required, given the individualistic and competitive nature of business interests, for owners of capital to be able to associate with each other and to form collective interest organizations?
3 Is the collective interest representation of employers subject to the same dialectical forces that have made workers' organizations, established initially to advance the interests of their constituents vis-à-vis other social groups, assume governing properties in relation to their members with time? (Schmitter et al., 1980)

With regard to the first question, the case-study of the chemical industry offers extensive evidence in favour of the thesis that capitalists, in spite of their structural advantage in market economies, cannot promote their interests without entering the political arena directly and that, in order to do so, they need various forms of associational action. Business and employer associations have historically been formed in order to cope with one or more of three basic problems that individual capitalists could not deal with by themselves. The first was the challenge presented by the political mobilization of workers to capitalists' discretionary power to invest and organize labour activities. The institutionalization of class conflict, the effort to guarantee a reliable labour supply and the setting of shared rules of the game in labour relations have become major concerns for business associations, and trade unions have become their most important counterpart. The second problem concerned the devastating consequences of cut-throat competition in the market, the access of foreign competitors to domestic markets, and the need to form a common front vis-à-vis the suppliers of the basic factors of production, such as bankers, sellers of raw materials, etc. The third problem concerned systematic increases in state intervention in the economy and in the regulation of economic processes by democratic political institutions which drew business more and more into the political game.

Our study of the chemical industry shows that business associability is a basic requirement even for a highly concentrated sector that enjoys a privileged status in the economy as an industry's industry, and in industrial relations due to its successful performance and its ability to offer higher than average wages. Individual firms, including the large oligopolies, need various forms of associational action to control overcapacity and regulate competition, to exert political influence over the formation of government policies, and to deal with unions and other major stakeholders.

The second and third questions concern the organizational properties that associations and associational systems assume in different environments in order to perform their institutional tasks. The aim of the research was to assess the ways in which, as well as the extent to which, associations are able to achieve a high degree of organizational development and, through collective action, reach their policy goals. These two dimensions – organizational development and policy achievement – are closely related. Associations that possess a developed organization, sufficient autonomy from their members and the state, sufficient authority over their members and a sufficient degree of legitimacy from the state may play an effective role in the political arena. In turn, demonstrated effectiveness in securing the implementation of agreements leads to the delegation of greater authority to an association, thus increasing its importance with respect to both its members and the state.

The organizational properties of business associations (formal domain of representation, internal structure, resources, tasks performed) were the specific object of our study. The major dependent variable was the composite concept of organizational development defined in terms of the two basic dimensions of organizational complexity and relative autonomy. The concept of organizational development in fact may be defined in light of main-stream organizational theory in which the degree of organizational development is measured in terms of the scope and purpose of the organization, the number and diversity of its units, the degree of specialization and functional coordination among its components, the extent of control over its environment, and its capacity to act autonomously and pursue long-term strategies. In contrast with the formulation of Schmitter and Streeck, developed organizations may be seen as open systems in which, according to systems theory, greater complexity goes hand in hand with a greater degree of structure through a systematic interchange between the organization and its environment. In this case the organization is able to extract more resources from its environment than it consumes.

The theoretical framework used to explain the problem of business associability and to account for the differences in the degree of organizational development in business associations was drawn from two basic logics of membership and influence. Organizations are in fact engaged in a double set of exchanges: with their members and with their interlocutors and relevant publics. Their organizational structures must often meet the conflicting needs of this double relationship which generally results in some sort of compromise. The logic of membership requires an articulated organizational structure that is capable of adapting to fragmented and segmented interests. The logic of influence, by contrast, requires an organizational structure that is capable of providing a unified representation of business interests vis-à-vis institutional counterparts such as unions, government agencies, consumer associations and environmental lobbies.

In spite of the fact that the above two logics are closely interrelated, it is possible to identify characteristics which are unique to each. The logic of membership refers to the social features of the members of a given association, such as the degree of internal social cohesion, and the awareness of common interests. It is based on the specific features of the economy, industrial sector, or subsector concerned, such as the degree of competition and diversification, the rate of growth, and the size of profits or losses. The logic of membership focuses on the kinds of questions typically asked by scholars of rational choice and collective action (i.e. Mancur Olson) and their critics who stress the importance of collective identity. The chapter by Chiesi in this volume discusses such themes as domain definition, boundaries, selective and solidaristic goods as incentives to membership, the management of diversity, and the influence of state intervention on the logic of membership. In Chapter 3 Grant then applies these ideas to an analysis of the specific characteristics of associations in the chemical industry in comparison with the other industries studied in the international project, both at the national and at the European level.

The logic of influence, on the other hand, is concerned with the various stakeholders of business interest associations. It looks outward to external influences and to the targets of associational action. These targets include national, regional and supranational governments, as well as trade unions. But they also encompass other components of associational systems such as competing business organizations, peak associations, chambers of commerce, large firms, political parties, special-interest groups, and consumer advocate associations. The chapters by Adams and by Whitston focus on the two major interlocutors of business organizations in the different advanced industrialized nations: the state and the trade unions. These chapters offer interesting insights into the assessment of the relative importance of national versus sectoral variables in explaining policies. The chapter by Jacek analyses the role of business associations as private interest governments in five major policy areas: the shaping of industrial structure (including the problem of overcapacity), the regulation of the labour market, the negotiation of binding agreements in the area of supplier–consumer relations, the regulation of product quality and health standards, and general macro- and intersectoral policies such as innovation policy, pollution control, etc. The chapters by Paterson and by Grant examine in detail two major industrial problems that are the object of both corporate strategies and public policies: environmental protection and industrial overcapacity in the chemical industry. Their findings concerning the role of business associations as agents of public policy vis-à-vis other institutional arrangements such as direct agreements among chemical producers and direct intervention of the state shed new light on some of the hypotheses of the research.

The concluding chapter by Martinelli and Grant proposes some significant modifications to the theoretical framework that acted as a basis for the international research project. These modifications concern both the relationship between organized interest politics and other forms of collective action by business, and the role of neocorporatist arrangements vis-à-vis other mechanisms of regulating the economy.

A general finding that has emerged from this study of the chemical industry is that business interest organizations are just one form of associational activity and just one of the ways through which business pursues and defends its interests and values. The increasing internationalization and concentration of the economy – which are features of the chemical industry, but which characterize other important industrial sectors as well – create the conditions for the growth of powerful corporations operating in a global market. These corporations, through sophisticated industrial and financial strategies (such as the organization of a tightly coordinated and diversified work process in different countries, the formation of mergers, acquisitions, and strategic alliances, etc.), may develop efficient plans at the international level, maintain oligopoly competition within tolerable limits, keep trade unions in check by organizing production in a number of different countries, and obtain a strong bargaining position vis-à-vis national government in order to influence relevant policy decisions. These corporations, as various chapters in this volume demonstrate, often consider business organizations a second-best strategy for the pursuit of their interests and the solution of their problems. Better strategies often include market action, direct influence on government representatives or agencies and negotiation with firms outside the associational structures. The management of the overcapacity problem in the chemical industry represents an excellent example. The management of this problem, in fact, has been achieved through negotiations between companies. In contrast to neocorporatist, tripartite forms of concertation, moreover, the state and international bodies played only marginal roles in this process.

This study also shows that the theoretical framework of the international research project overestimated both the role of business organizations in coping with company problems and prediction concerning a reliance on neocorporatist arrangements to regulate the economies of Western Europe. Using the 'asymmetry hypothesis' developed by Offe, one can argue that these two issues are closely linked (Offe, 1985). Labour and business have different attitudes towards organized collective action, business being less prone to neocorporatist arrangement than unions. Firms in fact have a variety of options for political action, besides organized collective activities through business associations: they can exert direct influence on the formulation of public policy through their operations in labour and product markets; they can enter into direct and individual negotiations with government agencies; and they can agree on price and output control through cartel action. Workers, on the other hand, can exert effective influence only through collective action in unions. The decline of neocorporatist arrangements in countries that were traditionally favourable to them (West Germany and the Scandinavian countries) and the failure of neocorporatist agreements in countries such as Italy are due to a considerable extent to the fact that tripartite concertation and neocorporatist agreements are generally a second-best solution for business.

The regulation of an advanced capitalist economy is a complex matter that is influenced by a complicated mix of institutional mechanisms. Interest organizations and neocorporatist, tripartite arrangements are just some elements of this mix. As argued above, Western Europe has applied a unique mix of institutional mechanisms to the governing of its economies, among which business organizations were only a part. This is particularly true for the chemical industry, where the strategic importance of the sector has induced governments to monitor and regulate closely the performance of the market, and where the oligopolistic structure of the industry grants large firms a larger role. Chapter 4 shows how the strategies of ICI and Montedison allowed these firms to act as agents of industrial policy both through direct relations with the state and through a soft and discrete but dominant role within sectoral associations. Grant's chapter on overcapacity management (Chapter 10) also shows the essential role of direct negotiations among large firms at the European level.

The choice of the chemical industry

The chemical industry provides an interesting testing ground for assessing the relative importance of business associations versus other institutional mechanisms in the governance of one of the most difficult and crucial areas of industrial restructuring. The chemical industry is a rather neglected area of study; certainly the amount of attention it has received from industrial economists and social scientists is not proportional to its importance in advanced market economies. In general, academics have tended to concentrate their research efforts on industries that have experienced dramatic declines in employment and output (i.e. steel) or on advanced technology industries that affect both the organization of industrial processes and the life of consumers. The chemical industry has undergone a crisis of overcapacity, due mostly to the oil price increases in the 1970s and the growth of competition from less developed countries. Chemicals is also a promising high-technology industry as demonstrated by the new innovations in the fields of biotechnology and new materials. Despite these characteristics, however, the chemical industry has not been perceived as either the most crisis-ridden or the most promising sector for study by scholars of industrial policy and government–business relations. This volume thus aims to fill an important gap in the literature.

But the choice of the chemical industry was also the outcome of a long selection process by the participants of the international comparative project. Emphasis was placed in the project on the analysis of associational action at the sectoral level, on the assumption that organized business's responses may differ substantially both across countries and across sectors. The literature on the role of business in politics is rich with references to the power of the capitalist class as a whole, and to the correlation between the strength of particular industrial sectors and specific public policy outcomes. It is very poor, however, in assessments of the mechanisms through which specific sectors of business develop an organized and conscious collective action for the pursuit of their collective sectoral goals. As Cawson points out, neither classes nor sectors as such act politically: class and sectoral interests are identified, promoted and defended through organization (Cawson, 1985).

After extensive discussion related to the relative importance of various sectors in the different countries participating in the project, we finally selected four industries (chemicals, food processing, machine tools and construction) to study on the basis of a sampling matrix. The matrix was constructed from two sets of variables: the economic structure of an industry, such as the degree of concentration of ownership and control, the type of technology, the degree of internationalization of production and sales, the degree of capital intensity, the number of relevant subsectors and the type of geographical distribution of firms; and the industry's relations with its major interlocutors and adversaries, such as the state (the extent to which an industry is subject to state ownership or to state regulation and is dependent on the state as a customer) and the unions (trade union membership, extent of militancy, ideologies, organizational structure). Each national team of the project collected a standard set of data on each of the sectors and subsectors chosen. The variables were specified in codebooks designed by the participating groups. Data, documentary information and interview material were also collected from published sources (such as the United Nations *International Standard Industrial Classification* and the European Community *General Classification of Economic Activities*), from public agencies and private research organizations, from business interest associations and trade unions, from social scientists and from individual firms in the nine countries. Further empirical evidence was collected for this volume from the Anglo-German research project on the chemical industry, from research on Montedison, and from other studies conducted by some of the contributors who were not part of the international research team. On the basis of this vast and heterogeneous material, the decision was made to separate industrial chemicals from pharmaceuticals. In this way we were able to identify the major characteristics of the chemical industry and of the collective organized activity action of this part of the business class.

The main characteristics of the chemical industry may be grouped under the following five headings:

Pervasiveness and economic centrality

The chemical industry is an industry's industry with both direct and indirect links with sectors such as textiles, health, agriculture, automobiles, housing and consumer durables. Davis argues that although the chemical industry in Western countries accounts for no more than 3–4 per cent of GNP and employs under 1 per cent of the labour force, it has a strategic impact on as much as 40 per cent of the economies of those countries (Davis, 1982). As Adams argues in Chapter 5 of this volume, moreover, in many countries chemicals have been treated not simply as a link between different sectors, but as a locomotive for growth. For all of these reasons the performance of the chemical industry is both vital for, and very much dependent on the general economic climate of the economy. Boom and slump periods in the various sectors of the chemical industry are closely related to the performance of other industries; the rubber and plastics industry depends heavily on trends in the automobile industry; fertilizers and pesticides depend on trends in the agro-business; artificial fibres depends on trends in the clothing and furniture industries.

Internationalization of trade, strategy and competition

A second distinctive feature of the chemical industry is its high level of internationalization both in terms of trade and in terms of production. As Mussati and Soru report in Chapter 2 not only countries with small domestic markets such as Switzerland, the Netherlands, and Belgium export a large percentage of their chemical production, but also larger countries such as the United Kingdom, West Germany and France. In 1986 foreign sales as a percentage of total sales were as much as 75 per cent for ICI, 72 per cent for Hoechst, 64 per cent for BASF and Bayer and 60 per cent for Rhone Poulenc. Moreover, a large percentage of the foreign sales of many of these European firms is actually produced in plants located in other European countries: 34 per cent for Hoechst and 37 per cent for Bayer. Finally in recent years there has been a substantial growth in both the sales and investments of European firms in the United States, including a number of acquisitions. Six European firms (ICI, Shell, Ciba-Geigy, Hoechst, Bayer and BASF) rank among the top 20 US companies on the basis of annual sales. By contrast, American companies began to increase their investments in Western Europe back in the 1960s in order to take advantage of the integrated European market.

Although US companies have also invested in other areas such as Latin America, and although Saudi Arabia's petrochemical industry is growing, most production and trade in chemicals takes place in the advanced industrialized countries. Six countries (USA, West Germany, Britain, France, Italy and Japan) in fact account for approximately three-fourths of total production.

Capital-intensive, extensive research and development

The concentration of the chemical industry in highly industrialized countries is related to a third distinctive feature of chemicals: the use of capital-intensive, continuous-production processes. Firms normally manufacture very large quantities of homogeneous materials, such as liquids and gases, which are best produced, processed and traded on a large scale. The chemical industry has also invested highly in research and development which has in turn resulted in significant innovations. It is quite common for major chemical firms to spend 3–4 per cent of sales of research, and these percentages may go as high as 10–15 per cent in specialty chemicals. According to OECD figures, moreover, over 20 per cent of the scientists and engineers working in the manufacturing sector are employed in chemicals. These requirements for high capital and research expenditures, and for large numbers of scientific and technical workers, explain in part why the world's chemical industry is largely confined to the advanced industrialized countries which have developed social, financial and intellectual infrastructures (Pettigrew, 1985). These factors, in fact, create high entry barriers that may only be overcome by firms that either have accumulated capital in other sectors such as oil, or are controlled by the state and receive large subsidies from the government.

Oligopolistic structure and markets

This accounts for a fourth distinctive feature of the industry: the oligopolistic structure of the markets for chemical products. The chemical industry has a long tradition of cartels and collusive action by firms. Cartels have recurrently been formed in fact to influence the determination of prices and production quotas at the international level. The German chemical monopoly was divided by the allied powers after the Second World War not only because of its role in the German war machine, but also to apply the new anti-trust policy. In the contemporary international economy, cartels have given way to inter-oligopoly competition among the major enterprises. These new trends, however, do not preclude strategic alliances, such as joint ventures, or successful strategies of rationalization, as in the case of the control of overcapacity.

Government involvement
The implications that developments in the chemical industry have for health and environmental issues, together with the economic importance and the capital- and research-intensive nature of the industry, go far to explain the extent of government involvement. State intervention has taken on different forms in the advanced industrialized nations, from public support for research in university centres or in public/private laboratories, to indirect fiscal and financial measures to spur innovation and increase competitiveness, to direct government intervention through state-controlled enterprises.

All of these features of the chemical industry have clear implications for associational action. First, the oligopolistic structure of the sector, dominated by new large corporations or a limited number of national champions, reduces the risk of fragmentation and the problem of organizing a multitude of different interests within a business association. As Chapter 3 shows, the chemical industry has a relatively coherent and well-organized associational system in which there are fewer first-order associations and fewer territorially specialized associations than in many other sectors, such as food processing and construction. Moreover, the structure of the chemical industry in which large and small firms coexist, raises the problem of managing the heterogeneity of interests which stem from differences in firm size. Various devices have been developed to cope with this problem in chemicals, including different financial contributions (which often imply a de facto subsidization of small firms by large ones), formal and informal rules of representation in executive boards for firms of different size, the granting of the right to dissent to minority groups, and the use of the association's selective goods.

Second, the capital-intensive character of the industry implies that salaries and wages contribute relatively little to total costs. This allows major corporations to accommodate the demands of the unions better than many other industries and guarantees relatively good labour relations, as Whitston demonstrates in Chapter 8.

Third, the international character of the industry and the rather specialized technical nature of the production process have contributed to the development of an international management culture among top managers. This facilitates concertation among these firms with regard to problems such as overcapacity on the one hand (see Chapter 9), and importance to supranational sectoral associations such as CEFIC on the other.

Finally the concerns of public authorities, unions and public opinion about the relationship of chemicals to the defence industry, the impact of chemicals production on the health of workers in the industry, and the importance of environmental and safety issues for consumers and local communities located near chemical plants, major transportation corridors, and disposal sites has fostered attempts by the chemical industry associations of various countries to deal with environmental hazards through collective self-regulation. In this respect the European experience of informal and cooperative relationships between the state and firms, and the role of associations as private interest governments seem more effective than the tough regulations adopted by the United States government and implemented by a variety of conflicting agencies, which have given rise to long and costly political and legal battles (Brickman et al., 1985; Vogel, 1986).

In light of the broad theoretical framework presented above concerning organizational development and the patterns of influence of business associations on public policy, and in light of the specific features of the chemical industry and their implications for business associations, we may identify nine major research questions which have guided the analysis in this volume:

1 Are business interest associations in the chemical industry viable instruments for the governance of industrial change and rationalization and for the resolution of problems such as excess capacity, or are other institutional actors (individual firms, state agencies, supranational bodies) likely to accomplish these tasks more effectively?
2 Is it in the government's interest to favour the organizational development and active policy role of associations that represent the collective interests of industry, or is it easier and more effective for governments to deal directly with the largest firms?
3 Are business associations in the chemical industry effective agents of public policy, and are they able to act as private interest governments?
4 Are the chemical industries in Western Europe able to cope with the conflicts that arise between the logic of membership and the logic of influence?
5 Have business interest associations in the chemical industry reached a significant level of organizational development in terms of domain, resources, structure, and outputs?
6 Do country styles of policy-making matter more than industry-specific problems in the setting of national industrial policies?
7 Will technological change and competitive pressures increase the importance of direct negotiations with workers at the firm or

plant level in the chemical industry, or will sectoral business associations and trade unions continue to play a major role?

8 Are business interest associations reliable institutions for setting the norms and standards of performance which must meet the environmental challenges that face the industry?

9 Are existing national business interest associations, working through European-level associations, a viable channel for the development of a European-wide coordinated strategy?

This volume attempts to provide satisfactory answers to these questions. It aims to make a significant and novel contribution to the analysis of industrial governance and change. We believe that this has been a field for too long dominated by economists and that it requires new inputs. We have adopted a political-economy approach in which economic issues are analysed by applying the theoretical constructs and research methods of political science and economic sociology. More specifically, we will not analyse the outcomes of industrial policies in terms of their economic results, but rather the formation of these policies and the role of organized business interests in light of a more general theory of interest intermediation and representation. The volume also aims to contribute to discussions on the organizational development and logics of business associations according to the model developed by the international project research team. We therefore hope to achieve a double result: to advance political and sociological theory in research on interest organizations and representation and in the debate concerning neocorporatism, and to offer a fresh perspective on pressing problems that until now have been analysed mainly according to traditional economic concepts.

References

Brickman, R., Jasanoaff, S. and Ilgen, T. (1985) *Controlling Chemicals: The Politics of Regulation in Europe and the United States*. Ithaca: Cornell University Press.

Cawson, A. (1985) *Organized Interests and the State: Studies in Meso-Corporatism*. London: Sage.

Chiesi, A. and Martinelli, A. (1989) 'The representation of business interests as a mechanism of social regulation', in P. Lange and M. Regini (eds), *State, Market and Social Regulation*. Cambridge, MA: Harvard University Press. pp. 187–213.

Davis, D.S. (1982) 'The chemical industry in its historical context', in D.H. Sharp and T.F. West (eds), *The Chemical Industry*. Chichester: Ellis Horwood.

Grant, W. (1987) *Business and Politics in Britain*. London: Macmillan Education.

Katzenstein, P. (1985) *Small States in World Markets*. Ithaca: Cornell University Press.

Offe, C. (1985) *Disorganized Capitalism: Contemporary Transformations of Work and Politics*. Cambridge: Polity Press.

Pettigrew, A. (1985) *The Awakening Giant: Continuity and Change in ICI*. Oxford: Basil Blackwell.

Schmitter, P. and Streeck, W. (1981) 'The organization of business interests: a research design to study the associative action of business in the advanced industrial societies of Western Europe'. Mimeo, International Institute of Management, Berlin.

Schmitter, P., Streeck, W. and Martinelli, A. (1980) 'The organization of business interests in advanced industrial societies – a project proposal'. Mimeo, International Institute of Management, Berlin.

Shonfield, A. (1965) *Modern Capitalism*. Oxford: Oxford University Press.

Vogel, D. (1986) *National Styles of Regulation: Environmental Policy in Great Britain and the United States*. Ithaca: Cornell University Press.

Zysman, J. (1983) *Governments, Markets and Growth*. Ithaca: Cornell University Press.

2

International Markets and Competitive Systems

Giuliano Mussati and Anna Soru

This chapter starts by outlining the main features of the industry we are dealing with, drawing attention to the historical origins of the different competitive models that characterize the modern chemical industry in the major industrialized countries. Then it summarizes the changes that have occurred in the competitive conditions, international performance of countries and firms' strategic behaviour in the past fifteen years. In these ways it provides a body of information about the industry which assists the understanding of the analyses provided in subsequent chapters.

Since 1973 it is possible to characterize two main periods of change in the international chemical industry: one, lasting from 1973 to 1982, covers the years of oil shocks and economic recession; the second one (1982 to 1990) is characterized by recovery from the industry's worst postwar recession. Firm strategies in the earlier period were aimed primarily at restructuring and diversifying production, while the emphasis in the later period was on improving competitive positions in core businesses.

At the beginning of the 1990s the competitive position of the industry is characterized by the following core elements:

1 growing involvement in high-technology chemicals;
2 progressive reduction in the number of actors;
3 ongoing research intended to produce technological leadership;
4 global growth in terms of production, trade and research.

New technological opportunities are now emerging, together with significant innovative advances in biotechnology and new materials. The United States maintains a technological lead in these fields, but the Japanese industry, traditionally not very competitive in the chemical sector, is becoming stronger.[1]

It must be emphasized, however, that the European industry has maintained a favourable performance in chemicals. West Germany and the United Kingdom, in particular, have registered a phase of growth characterized by an emphasis on internationalization and diversification, particularly in the US market, in order to strengthen their market position and develop cooperative agreements in relation to new technologies.

On the contrary, Italy, and to some extent France, have lagged behind the process of internationalization and innovation, and consequently the prospects for the domestic chemical industry in those countries are less favourable. Having completed an important process of restructuring, which has been particularly long and expensive, the Italian chemical industry has not been able to support adequate growth in high-technology sectors, nor a sufficient presence in international markets.[2]

The chemical industry in a comparative perspective

By classifying industrial sectors according to the level and type of technological development, it soon becomes evident that the chemical industry is essentially a science-based industry.[3] The technologies applied in the chemical industry have scientific roots, and industry growth has been closely linked to scientific discoveries. Unlike other sectors (except electronics), in chemicals the ability to conduct research internally and to exploit quickly the results of basic research done in public laboratories (of universities or other public research centres) is of crucial importance. In few other sectors does entrepreneurship, defined according to Schumpeter as the ability to transform inventions (scientific discoveries) into innovations (significant economic events), play such an important role.

It is important to note the interdisciplinary nature of research in the chemical industry. This feature has become more pronounced as trends in technological progress have made research in the chemical industry of greater interest to other scientific disciplines: biology in the case of biotechnology, physics in the case of new material.[4] Moreover, the changes in firm strategies that occurred in the 1970s placed a new emphasis on the requirements of consumer industries. As a result the competitive position of the sector has become more dependent on the ability to link itself with outside research opportunities.

Evidence for the highly research-intensive nature of the chemical industry is also provided by data on the ratio of research and development over sales which, in 1986, ranged between 4 per cent and 6 per cent in the United States, Germany, United Kingdom, France and Japan; only in Italy was this ratio lower (2.6 per cent) (see Table 2.1).

In all countries this ratio for chemicals is considerably higher

Table 2.1 *Research and development over sales (percentage)*

	1980	1981	1982	1983	1984	1985	1986
United States	2.8	3.1	3.8	3.8	3.8	4.1	4.4
Japan	3.1	3.4	3.7	4.0	4.2	4.6	5.1
West Germany	4.9	5.1	5.1	5.0	5.0	5.3	6.0
France	3.7	3.7	3.8	4.1	4.3	4.7	5.4
United Kingdom	3.4	3.7	3.9	4.0	4.0	3.9	4.0
Italy	1.3	1.6	2.1	2.1	1.9	2.2	2.6

Source: Our calculation based on *Federchimica* data.

than the average for other manufacturing industries. Moreover, the research and development expenditure of chemical firms as a percentage of the total research and development expenditure of all firms is particularly high in these countries.

The chemical industry, however, encompasses subsectors that are characterized by different levels of research intensity. Variations in the ratio of research and development expenditure over sales in the above-mentioned countries, therefore, reflect differences not only in the innovative strength, but also in the structure and composition of the national chemical industries.

The importance of research and development, and the opportunity to exploit the results of in-house research at international level, have resulted in an extensive internationalization of the chemical industry. This process began with commercial activities, but it has been extended into productive activities as well as cooperation agreements.

The West German chemical industry was able to use its innovative abilities at an early date in order to break the English monopoly in some natural colouring products such as 'indigo'. By 1910, in fact, the Germans were able to substitute these natural products and capture 80 per cent of the market for synthetic colouring.[5] The major German producers (Hoechst, Bayer, BASF, Agfa) also occupied important positions in areas such as pharmaceuticals, photosensitive materials and inorganic chemicals. In the period between the two world wars, IG Farben (created by a merger of Hoechst, Bayer, BASF, Agfa and some smaller companies) assumed a leading international position as a result of its worldwide commercial network, its ability to conduct research and its large financial resources. IG Farben was broken up after the Second World War. German firms, however, were only temporarily stopped in their conquest of foreign markets by the problems of reconstruction and restructuring which followed the war. They soon began to extend their commercial activity and, later, to make direct investments abroad, especially in Europe and Latin America (for example Brazil).

The existence of the British Empire and, later, the Commonwealth meant that the overseas activities of English chemical firms were directed mainly towards the Dominions (Australia, New Zealand, India, Canada, South Africa, etc.).

In the other major European nations (France, Belgium, the Netherlands), the respective leaders of the national chemical industries became the protagonists of internationalization, Rhone Poulenc, the French leader, exported 60 per cent of its total sales to foreign markets. Similarly, the Belgian firm, Solvay, produced only 13 per cent of its total sales within the Benelux nations. For countries such as Belgium and the Netherlands, in fact, internationalization was unavoidable because of the small size of the domestic market.

The same holds true for the Swiss chemical industry. The development of the Swiss industry shows the importance of research and the ability to maintain close relations with the market to achieve success under unfavourable conditions. The traditions of the Swiss chemical industry in terms of both production and research date back to the end of the nineteenth century. The eventual success of some of the large Swiss firms at the international level in the area of specialties led to a process of backward integration and to the production of intermediate goods.[6]

The level of internationalization of the Italian chemical industry was lower than that of its European competitors. By the beginning of the 1970s the main Italian firms produced abroad only a small percentage of total sales abroad: foreign averages for the period 1971–73 were: SNIA Viscosa 20 per cent; Montefibre 9–10 per cent; Carlo Erba 25–6 per cent; Montedison (excluding Montefibre and Carlo Erba) less than 5 per cent. Lepetit was the only firm able to reach the European level of 47–8 per cent (compared to ICI 54 per cent; BASF 46 per cent; Hoechst 55 per cent; Bayer 65 per cent).

Since then chemical firms have continued to grow internationally and the gap in the level of internationalization of Italian firms and non-Italian firms has increased: in 1986 the ratio of sales on foreign markets to total sales was about 25 per cent for Montedison and Enichem, compared to 75 per cent for ICI, 72 per cent for Hoechst and 64 per cent for BASF and Bayer.

For many of these European firms a substantial percentage of their foreign sales is actually produced in plants located in other European countries: ICI 22 per cent, Hoechst 34 per cent, BASF

Table 2.2 *The top six European chemical companies in the US (1987 estimates)*

	Sales ($bn)	Percentage of worldwide sales	US R&D expenditures as percentage of total
Hoechst	4.5	25	17
Bayer	4.0	20	23
BASF	3.8	19	19
Shell Chemicals	3.7	34	n.a.
ICI[a]	3.5	23	22
Ciba-Geigy	2.4	30	n.a.

[a] Including Stauffer Chemical.
n.a. not available.
Source: Company reports.

28 per cent, Bayer 37 per cent, Montedison 18 per cent. Recently, however, the weight of investments directed to the US market has grown considerably. About a dozen European chemical companies have annual sales in the US of $1 billion or more. Six of them rank among America's top twenty chemical companies on the basis of their American sales: Britain's ICI, Anglo-Dutch Shell Chemicals, Switzerland's Ciba-Geigy and West Germany's Hoechst, Bayer and BASF (see Table 2.2).

The large American firms long restricted themselves to the domestic market, where they were protected before the Second World War by tariff measures (American Selling Price), and to neighbouring markets such as Canada. Only later did they follow the European pattern, expanding into Latin America. During the 1960s the internationalization of the US chemical industry was expanded further with the growth of American direct investments abroad. In their period American firms began to locate productive activities in Europe (first in the United Kingdom, later in all European countries) in order to take advantage of the openness of these markets and the industrial infrastructures of these countries (Azzolini et al., 1979).

In Europe there has been a tendency to think in terms of national champions; in part for historical reasons, and in part as a result of government policies aimed at overcoming the difficulties of the first postwar period. In the 1930s the German and British governments, in fact, favoured respectively IG Farben and ICI in order to maximize international competitiveness. Until the end of the 1950s, moreover, the importance of major national chemical firms in the other European domestic markets increased considerably: Solvay in Belgium, Akzo in the Netherlands, Montecatini in Italy.

The focus on national markets has progressively declined, however, as the result of the creation of the EEC and the opening of the European market; foreign direct investment policies of the US firms in Europe; and the entrance of producers from other sectors into chemicals. The increasing internationalization of the chemical industry has also made the role of national markets less important for European firms. This is particularly true in France and Italy, where public companies and, more generally, public policies towards the chemical industry have traditionally played an important role.

Changes in the chemical industry during the 1970s

The chemical industry occupies a central position in the world economy. Its close links with other major industrial sectors, in fact, mean that there is a strong relationship between cyclical trends in the economy as a whole and growth in the chemical industry.

The changes that occurred in the global economy during the 1970s[7] had significant effects on the chemical industry. The most important of these changes included:

1 The rise in energy costs which led to a change in the structure of costs. Variable costs (proportional to the quantities produced) became a larger part of the total costs of petrochemical plants. In contrast with the previous decade, the construction of large plants did not result in a reduction of costs.
2 The diminished demand by user industries which resulted from the general economic recession, the decrease in substitution processes, the reduced use of petroleum products, and the increase in the imports of chemical products from developing countries. These factors resulted in a chronic overcapacity in the petrochemical companies of advanced industrialized countries and a decline of profits.
3 The appearance of new producers in developing countries, especially in petrochemicals, which threatened existing firms about the perspectives of the sector.
4 The upsurge of sensitivity to environmental and safety problems, which resulted in an increase in the costs of plants and equipment.

It is important to note that the changes that occurred during this period in industry resulted not only from a one-off increase in

Table 2.3 *Production indexes for chemicals and manufacturing: average annual rate of development (percentages)*

	1974–79	1979–82	1982–86	1974–86
Chemicals				
United States[a]	5.56	−2.16	7.10	4.08
Japan[a]	5.10	−1.47	4.43	3.20
West Germany	2.76	−4.02	2.61	2.01
France	3.09	0.08	1.22	1.71
United Kingdom[a]	1.43	−3.21	4.00	1.09
Italy	3.97	−0.35	2.51	2.39
Manufacturing				
United States[a]	3.35	−3.55	6.07	2.47
Japan[a]	3.36	2.79	4.84	3.71
West Germany	2.00	−1.97	3.02	1.33
France	1.47	−1.97	0.52	0.28
United Kingdom[a]	−0.36	−3.30	1.48	−0.50
Italy	1.87	−0.20	1.56	1.24

[a] Fibres are excluded from the 'Chemical' category.
Source: OCSE and CEFIC.

relative prices as a consequence of the first oil shock, but from continuous changes in prices with larger than normal fluctuations. The uncertainty that arose from these conditions had negative effects on the behaviour of firms and, in particular, on their investment strategies.

Before 1973 the world chemical industry was a leading industry with a growth rate in real terms superior to that of the economy as a whole. The growth of world chemical production was spurred by the growth of consumption that followed income trends, by the increase in demand by chemical user companies, and by technological advances that permitted the substitution of conventional materials with chemical products.

After 1973, however, this positive situation changed dramatically. The growth rate of the sector, which had previously reached exceptional levels, declined significantly, although it remained greater than that of manufacturing, with the exception of Japan and West Germany (see Table 2.3). This decline, moreover, created serious problems of surplus in productive capacity for bulk chemicals and resulted in low, or even negative, profit rates (see Ranci and Vaccà, 1979; Brancadore and Schieppati, 1983).

The strategies adopted by chemical (and in particular petrochemical) firms since 1973 have been constantly adapted to changes in the international environment (Mussati, 1983). The choices of firms, however, have not always been developed within the context of a coherent strategy, in part because neither the nature nor the importance of the changes that were occurring in the external environment were understood immediately by the firms.[8] The initial strategy of the petrochemical industry, in fact, was to attempt to cut costs by labour-saving and energy-saving techniques, and to integrate upstream with the petroleum industry. The general objective of these policies was to control the supply of feedstocks and to balance cracking capacities with lines of derivatives.

Only later did it become evident that the major problem was not so much the increase in the cost of raw materials, but the fall in demand that resulted from this increase. Declining demand made it difficult to transfer cost increases into the prices of final products. In addition, the change in demand for commodities in the chemical industry was of a structural nature, not a cyclical one. Even before the first oil shock, in fact, costs had begun to decline in the advanced industrialized countries.

Several factors, however, delayed adjustment, especially within Europe. Due to the extensive integration of intermediate and final products within the petrochemical industry, the elimination of productive capacity in one stage of production often implied changes in the entire production process, thereby increasing the costs of cuts (see *ECN Petrochemicals Supplement*, March 1985). Moreover, a large part of the surplus productive capacity had been built in the early 1970s. Soon after opening, therefore, many plants were confronted with the need to shift away from the production of commodities (which for many firms represented up to 80–90 per cent of total sales) to the production of specialties. Finally, many governments in Europe worried about the employment consequences of such a reconversion, and blocked any decision.

The strategies adopted by the chemical companies in the advanced industrialized nations to cope with the new conditions of the world economy varied significantly (Bower, 1986; Mansfield, 1986) (see Tables 2.3–2.5). In the United Kingdom and in Japan the industry experienced a sharp decline in production and a decrease in employment during the 1970s. This policy was continued during the short recovery of the 1980s, resulting in a 30 per cent reduction in employment over the period 1974–82. In these two countries, production and employment cuts particularly affected basic sectors, and were associated with considerable investments in specialties,[9] which attempted to capitalize on production and technological niches open to individual firms rather than reorienting them.

Table 2.4 *Investments on sales in the chemical industry (percentages)*

	1974	1975	1976	1977	1978	1979	1980	1981	1982	1983	1984	1985	1986	1987
United States[a]	6.97	6.92	6.59	6.00	5.61	7.16	7.92	7.77	7.68	6.97	7.19	7.59	8.47	8.06
Japan[a,b]	8.42	9.61	7.05	5.96	5.19	5.29	6.22	6.21	6.39	6.22	6.09	6.93	7.00	7.90
Total EEC[c]	6.52	8.04	6.72	6.71	6.43	5.72	5.94	5.36	4.61	4.67	4.50	4.65	5.63	5.91
West Germany	6.59	7.09	6.52	6.30	6.28	5.86	6.40	6.16	5.94	5.52	4.97	4.79	6.22	6.40
France[a]	5.37	6.28	4.67	4.34	4.40	4.40	4.25	3.81	3.12	3.10	3.50	4.09	5.01	5.35
United Kingdom[a]	4.01	6.18	5.50	5.94	7.64	6.99	7.86	8.07	4.36	5.29	5.43	6.26	6.43	6.14
Italy	10.88	15.06	10.00	10.53	7.24	7.01	6.75	4.70	4.82	4.60	4.09	3.73	4.27	4.83
Netherlands	6.15	7.76	8.89	11.43	8.46	4.65	4.70	4.11	5.05	5.04	5.92	5.43	7.20	8.38
Belgium[a,d]	4.25	8.87	7.46	4.96	5.00	3.62	3.18	2.38	2.44	2.87	2.67	3.15	4.16	4.57

[a] Fibres excluded.
[b] Includes rubber and plastic materials.
[c] Greece excluded.
[d] Permanent employment: data non-comparable.
Source: CEFIC.

Table 2.5 *Employment in the chemical industry (thousands)*

	1974	1975	1976	1977	1978	1979	1980	1981	1982	1983	1984	1985	1986	1987
United States[a]	1043	1013	1034	1057	1088	1113	1113	1108	1079	1058	1061	1045	1027	1024
Japan[a,b]	551	472	449	429	414	410	410	409	405	402	400	395	396	391
Total EEC[c]	1900	1835	1800	1815	1814	1796	1781	1742	1659	1622	1612	1604	1620	1618
West Germany	591	586	556	553	548	547	550	565	559	549	550	557	567	572
France[a]	310	304	304	307	306	301	296	289	280	282	279	272	269	264
United Kingdom[a]	482	428	423	429	439	441	431	403	348	330	329	336	336	337
Italy	298	297	296	293	292	291	283	271	259	251	241	233	230	225
Netherlands	93	98	98	96	94	92	91	91	90	88	89	89	91	91
Belgium[a,d]	93	91	91	92	89	89	90	88	88	88	89	90	91	91

[a] Fibres excluded.
[b] Includes rubber and plastic materials.
[c] Greece excluded.
[d] Permanent employment: data non-comparable.
Source: CEFIC.

Chemical companies in France and Italy, by contrast, delayed adjustments, initially simply cutting back investments. Not until the 1980s, in fact, were significant reductions in employment achieved in these two countries.[10] In the 1980s Italy and France too have achieved cuts in production and employment (especially in Italy); however, they have fallen behind other competitors in the manufacturing diversification process. The low level of investments, and research too in Italy, have not permitted the proper development of specialty products.

In Belgium and in the Netherlands supply has not yet been adjusted to the fall in demand. Only limited reductions in production and employment have been made, while investments have returned to pre-crisis levels following a short period of growth. It is important to note, however, that the foreign direct investments of American and German firms play an important role in these two countries.

Finally, in the United States and in Germany, total employment in the chemical industry has only slightly decreased while production has maintained high growth rates. In these two countries, overcapacity in the petrochemical industry was less serious than elsewhere as petrochemicals accounted for a smaller proportion of total output than in Japan and the other European countries, and the strengthening of high value-added sectors offset cuts in commodity products. Moreover, the American chemical industry was able to exploit the advantages of modern and efficient plants spread throughout the world and of large national energy supplies (especially natural gas) to keep costs down. The American industry was also able to take advantage of its worldwide flexibility, which was rooted in the internationalization policies of the 1960s.

In general, differences in technological capacity, the availability of natural resources and structural relations with the market account for the variations in the strategies that were adopted by firms to cope with the new global economic environment. The chemical divisions of the large petrochemical multinationals (such as Exxon and Shell) strengthened their positions of leadership in bulk petrochemicals by exploiting advantages in terms of raw materials and productive plants. According to the estimates of Arthur D. Little (1986) these divisions grew at an average annual rate of 27 per cent between 1974 and 1979 as a result of the effects of both the increase in petroleum prices and the growth of investments in petrochemicals by oil multinationals. This latter phenomenon was favoured by the large cash flows that these multinationals controlled during the 1970s, and by the limited opportunities for them to reinvest in the energy sector.

Table 2.6 *Classification of chemical products*

	Undifferentiated	Differentiated
High volumes	Commodities	Semi-commodities
Low volumes	Fine chemicals	Specialty chemicals

Other firms in the commodity sector sought to reduce their vulnerability with respect to the availability and cost of raw materials in two ways. The first consisted of upstream integration into the oil sector through direct investments or joint ventures (principally with American or European firms), or through the exchange of technology for raw materials (principally with Japanese firms that received government assistance). The second strategy, followed by many large European and American firms, consisted of diversification into the production of specialties or other high value-added products. This strategy was dictated in part by the investment projects of Canada, Mexico and Saudi Arabia in the petrochemical industry.

Trends in international trade in the 1980s

As noted above, the various subsectors of the chemical industry are extremely heterogeneous. These subsectors are normally classified according to the phase of the product cycle to which they belong (such as primary and secondary chemicals). Alternatively, these subsectors may be classified according to level of product differentiation and volume produced. In this case we may distinguish between commodities (undifferentiated products, high-volume production) and specialties (differentiated products, low-volume production) (Kline, 1972). This classification is best explained by Table 2.6. It is interesting to note that the importance of economies of scale and process innovations increases as one moves from right to left and from bottom to top in the table. In the opposite direction (left to right, top to bottom), the importance of product innovations and research applications, as well as the intensity of research and marketing activities, increases. In the case of commodities, in fact, engineering plays a crucial role while in the case of specialties, chemistry and biochemistry are critical.

After 1973 the industrialized countries' share of world chemical export declined at a rate that was similar to that registered for manufacturing industry as a whole. This shift favoured developing, non-oil-rich and Eastern bloc countries but not developing oil-rich countries. Only recently has the penetration of OPEC countries

Table 2.7 *Comparison of petrochemical imports into major world trading blocs – August 1986 (tons)*

	EEC		Japan		United States	
Products	Imports from Saudi Arabia	Extra-EEC imports	Imports from Saudi Arabia	Total imports	Imports from Saudi Arabia	Total imports
Styrene	23,782	123,064	86,424	171,535	0	73,107
EDC	2,020	18,637	170,464	390,310	0	22,767
Methanol	401,528	1,452,799	221,373	722,226	37,840	749,726
MEG	34,748	81,771	82,364	165,933	15,856	107,766
DEG	12,796	28,415	10,285	15,141	0	13,466
IIDPE	137,584	235,735	10,718	76,503	14,153	225,938
hdPE	50,253	172,396	3,455	12,343	4,244	17,410

Sources: EUROSTAT and national statistics.

(particularly of Saudi Arabia) in Western markets (mainly in the EEC and Japan) become important, and this is true only for a few bulk productions (see Table 2.7).[11]

A number of observations may be made from Table 2.8 concerning the market shares of the advanced industrialized nations. First, despite a slight loss of market share during the 1970s, West Germany remains the major exporting country. The German chemical industry, moreover, appears to have entered a new era of growth: its trade surplus (approximately $18 billion in 1987) has generally exceeded that of the American industry (see Table 2.9).

Second, despite the fact that the American industry was able to strengthen its competitive position after the two oil shocks as a result of relatively low energy costs, the US chemical trade was penalized in the first half of the 1980s by the appreciation of the dollar. In contrast with other manufacturing industries, however, trade in the US chemical industry has maintained a positive balance, with a surplus of almost £10 billion in 1987.[12]

Third, the trade surpluses of France and the United Kingdom are also relatively high. The French industry has slowly, but consistently, improved its trade balance as a result of decreases in imports and exports. These figures indicate that the massive restructuring of the French chemical industry towards high-technology sectors in the 1970s is finally beginning to foster positive results. Although significant, the market share of the British industry has declined slightly since 1979. This may still be considered as a positive result, however, if compared with the

Table 2.8 *Chemical industry: export market shares (percentages) on total OECD (OECD share = 100)*

	1971	1972	1973	1974	1975	1976	1977	1978	1979	1980	1981	1982	1983	1984	1985	1986	1987[a]
United States	18.20	16.31	16.27	15.73	16.53	16.52	15.85	16.25	17.04	18.29	19.82	18.13	17.42	18.45	16.92	14.86	14.24
Japan	7.06	7.04	6.06	7.26	7.38	6.22	6.29	6.07	5.54	5.40	5.77	5.68	6.01	6.14	6.00	6.25	6.38
West Germany	21.52	21.37	22.52	21.64	18.94	20.80	20.81	22.72	20.67	19.83	18.94	19.30	19.60	19.26	19.31	20.89	21.02
France	9.30	9.58	9.99	8.95	10.27	10.08	10.48	10.66	10.74	10.80	10.47	10.66	10.45	10.36	10.92	11.22	11.30
United Kingdom	10.22	9.49	8.84	8.97	9.17	9.08	9.89	9.73	9.64	10.03	9.34	9.73	9.26	9.08	9.69	9.51	9.57
Italy	4.98	4.82	4.68	5.58	5.09	4.78	4.62	4.47	5.36	4.49	4.50	4.82	5.40	5.11	5.26	4.73	4.74

[a] Estimated.
Source: OECD.

Table 2.9 *Chemical industry: trade balances (millions of dollars)*

	1974	1975	1976	1977	1978	1979	1980	1981	1982	1983	1984	1985	1986	1987
United States	4,967	4,699	4,933	5,566	5,840	9,829	12,211	11,870	1,045	9,040	10,200	7,230	7,814	10,168
Japan	2,310	1,832	1,084	1,296	1,339	922	565	340	−438	−223	−728	−406	−273	−194
West Germany	7,663	5,891	6,800	7,600	8,874	10,858	11,447	10,536	9,832	10,077	11,122	11,083	14,344	17,625
France	465	856	669	1,185	1,555	2,625	2,565	2,650	2,090	2,200	2,985	2,761	3,133	3,581
United Kingdom	1,741	1,786	1,888	2,444	2,769	3,204	4,983	4,562	3,394	2,393	2,164	3,067	3,355	3,512
Italy	−242	−271	−962	−647	−1,494	−2,750	−3,081	−1,596	−2,447	−2,295	−2,776	−2,726	−4,306	−5,384

Source: OECD.

performance of other manufacturing industries in the United Kingdom.

Finally, the two countries that registered negative trade balances were Japan and Italy. In the Japanese case there has been an improvement of the trade balance. In 1987 it was only slightly negative, despite the weakness of the Japanese chemical industry. In contrast with other manufacturing industries such as mechanics and transportation, in fact, chemicals still play a marginal role in the international competitiveness of Japan.

Italy has not been able to improve its competitiveness in chemical trade. Its share of the world market has remained consistently low and, since the 1960s, its trade balance has been negative. Italy, in fact, exports only approximately 25 per cent of total sales in chemicals; this figure is approximately 40 per cent in France and the United Kingdom, and greater than 50 per cent in West Germany (see Table 2.10). The low exports/sales ratio in Italy, moreover, has not been compensated (as in Japan and the US) by a low level of import penetration. The ratio of imports on apparent consumption, in fact, has increased from 25 per cent in the 1970s to 34 per cent in 1987, a level similar to that of its European neighbours (see Table 2.11).

If we examine the evolution of exports markets (see Table 2.12), it is interesting to note the relatively greater importance of the markets of industrialized countries (more sophisticated and more competitive for the presence of technologically advanced firms) for the large European producers than for Japan or the US. The importance of American and European markets has increased for all countries, while the importance of the markets of less developed countries and of centrally planned economies has decreased.

It is likely that the productive and export structures of the advanced industrialized nations will register a shift away from commodities and towards specialties. This shift will entail increased specialization in low-volume products, especially in special compounds that are generally characterized by greater product differentiation and higher intensity of marketing and research activities.

An analysis of the structure of chemical exports in the OECD countries demonstrates, in fact, that between 1981 and 1986 there was a shift from commodities (whose weight diminished from 30.89 to 26.10) to other types of product (+2.2 per cent for specialties; +1.33 for fine chemicals; +1.17 per cent for pseudo-commodities).

All countries analysed here, with the exception of the US, have recorded a decline in the export of commodities and an increase in

Table 2.10 *Chemical industry: exports on sales (percentages)*

	1974	1975	1976	1977	1978	1979	1980	1981	1982	1983	1984	1985	1986	1987
United States[a]	12.30	9.54	9.82	9.49	9.97	11.60	12.85	12.10	11.51	10.44	10.51	10.16	11.50	12.30
Japan[a]	15.12	11.61	9.38	9.31	8.35	8.93	9.00	8.32	8.56	8.63	8.73	9.00	8.35	8.56
Total EEC[b]	40.89	31.07	39.62	39.74	40.75	41.49	41.40	44.66	45.10	47.51	49.41	50.42	50.32	50.52
West Germany	44.10	36.33	41.01	41.21	41.27	42.96	43.94	46.19	47.21	48.34	50.73	51.71	51.45	51.52
France[a]	29.72	29.25	32.10	30.64	33.31	35.01	34.79	35.61	35.84	37.04	41.47	40.16	39.67	40.50
United Kingdom[a]	28.68	26.01	26.41	27.97	29.23	29.28	29.97	41.84	37.83	40.93	42.71	46.43	46.84	45.96
Italy	23.94	21.50	21.33	23.28	21.80	21.86	20.48	23.20	21.94	24.71	24.87	26.93	25.01	24.73
Netherlands	84.34	84.52	86.79	85.97	85.82	87.75	90.71	88.01	90.20	91.94	92.89	92.59	97.99	97.47
Belgium[a,c]	77.32	80.42	74.82	67.58	71.67	64.29	63.09	59.82	62.70	66.60	67.26	68.04	70.87	74.28

[a] Fibres excluded.
[b] Greece excluded.
[c] Includes rubber and plastic materials.
Source: CEFIC.

Table 2.11 *Chemical industry: imports on apparent consumption*

	1974	1975	1976	1977	1978	1979	1980	1981	1982	1983	1984	1985	1986	1987
United States[a]	6.62	4.58	5.21	4.84	5.62	5.37	5.72	5.71	nd	5.94	6.76	7.64	7.87	7.93
Japan[a]	9.04	6.50	6.85	6.69	6.30	7.68	8.31	7.94	9.10	8.88	9.49	9.43	8.57	8.69
Total EEC[b]	31.80	26.81	31.49	31.45	32.54	33.95	33.70	36.39	37.41	39.64	41.67	42.92	43.19	43.43
West Germany	26.66	22.02	25.99	26.56	26.67	29.23	30.50	32.37	33.79	34.62	36.60	38.14	37.56	37.40
France[a]	27.91	25.84	29.65	26.61	28.84	29.38	29.61	30.04	31.24	31.96	33.11	34.47	34.47	35.59
United Kingdom[a]	22.10	18.62	19.06	19.86	21.33	22.29	20.30	31.74	29.36	31.68	38.37	39.35	40.16	40.34
Italy	22.24	23.03	26.25	26.44	27.82	29.98	28.18	27.93	29.54	30.80	32.99	34.11	34.24	34.18
Netherlands	76.02	75.80	79.47	78.36	78.36	81.02	85.58	81.34	84.53	86.64	88.30	88.63	96.73	95.88
Belgium[a,c]	69.56	74.20	65.84	61.40	63.63	56.18	54.99	51.13	54.58	58.82	59.61	60.47	64.68	67.69

[a] Fibres excluded.
[b] Greece excluded.
[c] Includes rubber and plastic materials.
Source: CEFIC.

Table 2.12 *Composition of export markets by areas (percentages)*

		OECD Total	OECD Europe	EEC (9)	Others EEC	EFTA	Planned economies	OPEC	LDC	United States	Others
United States	1981	0.52	0.25	0.22	0.01	0.03	0.01	0.07	0.44	—	0.13
	1986	0.59	0.31	0.26	0.01	0.03	0.02	0.05	0.38	—	0.16
Japan	1981	0.31	0.11	0.10	0.01	0.01	0.05	0.00	0.61	0.14	0.02
	1986	0.39	0.17	0.14	0.01	0.02	0.04	0.05	0.55	0.19	0.01
West Germany	1981	0.73	0.59	0.45	0.04	0.15	0.06	0.05	0.19	0.05	0.05
	1986	0.78	0.67	0.47	0.04	0.15	0.05	0.03	0.15	0.06	0.05
France	1981	0.71	0.58	0.50	0.05	0.08	0.06	0.06	0.22	0.05	0.03
	1986	0.77	0.65	0.52	0.05	0.08	0.03	0.05	0.19	0.07	0.04
United Kingdom	1981	0.68	0.56	0.44	0.03	0.09	0.03	0.11	0.24	0.06	0.06
	1986	0.74	0.58	0.46	0.04	0.08	0.03	0.07	0.20	0.09	0.06
Italy	1981	0.56	0.45	0.33	0.06	0.08	0.06	0.15	0.35	0.05	0.03
	1986	0.68	0.58	0.40	0.07	0.09	0.05	0.07	0.25	0.07	0.03
Total OECD	1981	0.66	0.51	0.40	0.03	0.08	0.05	0.07	0.27	0.06	0.06
	1986	0.72	0.56	0.44	0.03	0.08	0.04	0.05	0.22	0.08	0.06

Source: OECD.

specialties. This shift in export structure has been most pronounced in Japan and Italy, the only two countries with a negative trade balance in specialties (see Table 2.13).

In the case of Japan the evolution towards specialized production is the result of a political agreement designed by MITI in the 1970s to identify priority areas within the chemical industry. In the case of Italy it is the result of the process of deindustrialization in the petrochemical sector which began in the late 1970s, and failed, despite efforts to boost specialties production, to lead to an improvement of the trade balance. All other countries present a positive trade balance in specialties, but only the United Kingdom presents a growing trend. The British industry, however, is the only industry besides the Italian that has registered a negative trade balance in both commodities and pseudo-commodities.

Looking across various models of export performance in the countries analysed, we find some variations. West Germany has significant market shares in all the large subsectors of the chemical industry and in each case these shares have grown from their 1981 levels. The French industry has slightly improved its export performance in commodities, in which its absolute level of specialization is higher, and has maintained its position in other subsectors. Finally, the US industry has experienced a high reduction of exports in all chemical sectors due to the devaluation of the dollar. This decline has been most marked in specialty products.

New strategies in the world chemical industry

To better characterize changes in firm strategy between the 1970s and 1980s, use will be made of Porter's methodology on strategic groups. A strategic group means a group of companies operating in the same sector following similar strategies in terms of the main decision-making variables (Porter, 1979). These firms tend to make similar responses to external stimuli, and are aware of the interdependence of their strategic choices (Caves and Porter, 1977). The fact that a company belongs to a given group ensures that it will not be imitated by companies outside the group without substantial costs, significant elapsed time, or uncertainty about the outcome of those decisions (McGee and Thomas, 1986).

In the 1970s for companies with a strong presence in petrochemicals the main variables in detecting strategic groups were fundamentally linked with cost leadership factors (in fact, technology was mainly process technology and easily transferable by licences), while firms operating in specialty markets relied mainly on scientific knowledge and technological factors. In the

Table 2.13 *Comparison between the structure and the performance of main countries on the basis of product classes*

Countries	Commodities		Pseudo-commodities		Fine chemicals		Specialties	
	1981	1986	1981	1986	1981	1986	1981	1986
United States								
percentage on exports	32.35	29.63	5.11	5.98	22.21	28.10	40.03	35.99
percentage on imports	36.87	26.88	3.65	4.96	37.83	37.25	21.14	29.86
as percentage of imports	0.88	1.10	1.40	1.21	0.59	0.75	1.89	1.21
market share	20.76	16.87	15.74	11.70	17.05	15.38	22.35	14.19
Japan								
percentage on exports	29.89	22.06	10.01	12.30	30.15	31.19	29.28	33.71
percentage on imports	21.01	17.35	4.18	4.58	36.43	37.78	37.92	39.64
as percentage of imports	1.42	1.27	2.39	2.69	0.83	0.83	0.77	0.85
market share	5.59	5.28	8.98	10.11	6.74	7.17	4.76	5.59
West Germany								
percentage on exports	25.26	21.74	8.08	8.96	27.20	27.10	38.37	41.19
percentage on imports	33.45	27.47	8.31	9.67	30.25	28.79	27.37	33.36
as percentage of imports	0.76	0.79	0.97	0.93	0.90	0.94	1.40	1.23
market share	15.49	17.40	23.80	24.62	19.96	20.85	20.47	22.83
France								
percentage on exports	21.38	27.90	5.92	6.40	25.20	25.20	36.89	39.86
percentage on imports	28.42	26.99	6.39	7.46	37.04	31.64	27.30	33.03
as percentage of imports	0.75	1.03	0.93	0.86	0.68	0.80	1.35	1.21
market share	10.63	12.01	9.63	9.45	10.22	10.41	10.88	11.87
United Kingdom								
percentage on exports	19.93	14.49	5.57	5.61	30.36	33.84	43.11	45.20
percentage on imports	27.30	26.93	8.43	9.18	31.02	23.28	32.71	37.82
as percentage of imports	0.73	0.54	0.66	0.61	0.98	1.45	1.32	1.20
market share	6.02	5.28	8.08	78.02	10.98	11.85	11.34	11.40
Italy								
percentage on exports	34.38	25.21	7.26	9.80	25.42	24.67	31.40	39.10
percentage on imports	30.54	26.37	7.47	8.98	26.17	22.79	35.01	41.13
as percentage of imports	1.13	0.96	0.97	1.09	0.97	1.08	0.90	0.95
market share	5.01	4.57	5.07	6.10	4.43	4.29	3.98	4.90

Source: OECD.

1980s market globalization, technological progress acceleration, macroeconomic climate deterioration (instability of exchange and interest rates) and the oil price decline have modified the boundaries between strategic groups, altering entry barriers which are now increasingly technological. Consequently, the strategies adopted by companies have changed, in general according to the following guidelines (see Ergas, 1984):

1 Strengthening value-added sectors and technological integration of various activities which, although diversified, are hinged on a common technological base which is itself strengthened. The considerable growth of firms' R&D investments has played a strategically important role.
2 Market segmentation by using appropriate technologies to ensure differentiated production.
3 Manufacturing focused on core businesses.
4 An emphasis on international growth.

A great deal of the money that the big chemical companies have raised through disinvestments in bulk chemicals, and saved through rationalization, has been invested into extra R&D. According to Kline (1972) and to Table 2.1, spending on R&D by American chemical firms increased from 2.5 per cent of sales in 1979 to 4.4 per cent in 1986. These funds have also been spent on acquiring specialty chemical firms, especially in the United States.

In the late 1980s more than 400 specialty chemical firms were sold in the US, a doubling of the equivalent number in the early 1980s. ICI (United Kingdom), for example, acquired Beatrice Co., thereby expanding its interests in specialties production. In the US Allied Co. (US) bought Bendix Co. in an attempt to enter the market for diagnostic and electronic chemicals. Finally, Hoechst (Germany) acquired Celanese (US), a manufacturer of specialty synthetic fibres.

Acquisitions of business lines or parts of companies also tend to involve the major chemical firms in the US and Western Europe (see Table 2.14). It is not uncommon for a company to be both a buyer and a seller, in an attempt to improve its competitive position in core businesses while shedding assets that have little potential for future growth.

It is necessary to point out that focusing strategies do not rule out inter-sector and inter-industry diversification strategies. In contrast to the 1970s, such choices of strategy are not imposed by an objective of risk minimization,[13] or by indiscriminate entry into new sectors, but they aim at exploiting complementarities between technological and manufacturing strengths.

Table 2.14 *Number of main acquisitions in the chemical industry (1985–87)*

Countries of origin	Areas of destination: Europe	North America	Japan	Others	Total	Percentage
United States	52	4	2	1	59	14.2
Japan	4	2	—	—	6	1.4
West Germany	44	25	1	3	73	17.5
France	13	21	2	6	42	10.1
United Kingdom	32	58	—	11	101	24.3
Italy	6	11	—	1	18	4.3
Netherlands	21	11	1	1	34	8.2
Switzerland	7	9	2	—	18	4.3
Sweden	15	5	—	1	21	5.0
Others	30	11	1	2	44	10.6
Total	224	157	9	26	416	100.0
Percentage	53.8	37.7	2.2	6.3	100.0	

Source: our elaborations on data from *European Chemical News* and *Mergers & Acquisitions*.

Diversification strategies have been differently developed in Europe and the US. In Europe the limited size of national markets has hampered the creation of large specialized companies: for example, in the pharmaceutical sector there are new large European companies started as pharmaceutical companies and mainly operating in this sector. Markets are dominated by large companies which expanded their activity from basic sectors to related ones.

In the US, however, high levels of demand have made possible the creation of large specialized companies, and diversification choices are mainly oriented to grasp technological and market opportunities in correlated sectors (from a technological point of view). For example, diversification choices in biotechnologies, electronics and aerospace have been numerous (see Mariotti, 1986; Cainarca and Mariotti, 1985).

In the 1980s the average annual growth rate of chemical production was higher in Japan and North America than in Europe. The growth rate of specialties production was higher than that of commodities.

By the beginning of the 1980s the largest market for commodities within the developed and non-planned economies (including OPEC and NICs) was Western Europe (35 per cent), followed by North

America (approximately 30 per cent). The opposite was true, however, for specialties: North America represented 40 per cent of the market, while Western Europe represented only 33 per cent.

Given that the US represents the largest market in absolute terms, and the fastest-growing market for specialties, many European firms took advantage of the favourable exchange rates in the second half of the 1970s to undertake direct investments in the US; their major objective was to acquire specialties firms. Between 1974 and 1979, in fact, it is estimated that the major European firms made more than fifty relevant acquisitions of this type. This trend continued even after the favourable terms of exchange had disappeared, indicating the strategic nature of the European decision to enter specialties production. The North American market is the most advanced in the world. In contrast with the European market, moreover, it is a much more homogeneous market. For firms that operate in a 'global industry'[14] (according to Porter's definition), as chemicals are increasingly becoming, a presence in the North American market therefore represents an important point of reference, and an obligatory step in the move towards internationalization.

In both Italy and France the move to specialties production was favoured by government policies towards the restructuring of the chemical industry. Rhone Poulenc (France) and Montedison (Italy) began to concentrate their efforts in high value-added sectors and to sell their primary chemicals plants to the nationalized oil companies. Similarly in Japan, MITI induced chemical firms to focus on biotechnology, electronic chemicals and carbon fibres for aeronautics.

It is important to note that almost 10 per cent of the chemical industry in the major European nations is controlled by the public sector; this figure is even larger in Italy and France, where it exceeds 50 per cent. With the exception of the French firm SNPE, these firms operate almost exclusively in petrochemicals, although the acquisition policies that they adopted in the recent past often led to a change in their products mix, moving away from commodities towards specialties.

Current restructuring represents a shift from diversification towards improving competitive position in core businesses. After the recent drop in oil prices, chemical firms have been induced to revise their strategies. The reallocation of production in the petrochemical industry towards the oil-rich countries has been, for the most part, completed. This process, therefore, should not be affected in the medium to long term, especially since prices should eventually be stabilized at a level slightly higher than those of the 1980s. It is true that the competitive advantages of these countries in the petrochemical industry are concentrated in variable costs, and that these should decline to less than 50 per cent of total costs in petrochemicals. If we exclude the projects that are aimed at internal needs (mostly in the planned economies and Mexico), most new plants (especially in the Middle East) have been completed. This might lead governments to maintain production, thereby leading to a possible price war. But even if the effects of the appearance of new producers on international prices are contained by their American and Japanese partners, they will result in a tension in global markets and the position of European producers in the international market will inevitably be weakened.

Conclusions

The world chemical industry, after a long period of crisis, experienced improved conditions in the late 1980s and many observers suggest that the sector has entered a new era, perhaps even a golden age.

Notwithstanding the current favourable trend, the chemical industry faces several problems including an increasing maturity of many of its products, intensification of global competition, overcapacity and fewer growth niches. As a result the global chemical business is now unlikely to recapture the rapid growth that characterized the 1960s and early 1970s. Despite these problems, new technological developments continue to move forward the chemical industry. The functional boundaries that in the past have isolated different sciences are now giving way to a truly interdisciplinary industry that will combine the resources of chemistry, physics, biology and electronics. So it is possible to identify many specific areas that show high potentiality for growth. *Opportunities in Chemistry*, the report published by the National Research Academy of Sciences in November 1985, outlines the potential advances in the chemical industry and identifies some primary areas for research and technological development for the next twenty years. The exploitation of these potential advantages will require the development of new skills that will be able to manage new technologies and aim at a competitive edge. On the contrary, in the mature segments of the chemical industry, competition and markets are often global. They require a new set of strategies that are able to manage resources and information on a global basis and to integrate structures for purchasing, converting and marketing products around the world in such a way that the lowest costs and highest values are achieved.

The long-term strategies of chemical firms will also be influenced by the decreasing rate of growth in consumer goods markets (at least in the industrialized countries) and by the existence of fewer growth niches, in this age of high availability of technology, which do not quickly become overcrowded with competitors.

Companies are reacting to mature markets in many different ways:

1 by tailoring new technologies to gain an edge in defined market niches;
2 by improving sales distribution systems and supplying back-up services to improve the added value of sales;
3 by accepting the reality of replacement markets and by treating them as an opportunity (that is, joint ventures between Japan and US firms and Arab ones);
4 by selling businesses in the hope of being able to reinvest the cash more profitably;
5 by adopting new strategic options directed fundamentally to specialization and internationalization.

In this way the firms hope to maintain high rates of growth even if their targets are very often inconsistent with the real opportunities of the market.

A survey made by SRI International (1985) shows that the firms interviewed have very high growth targets (rarely below 10 per cent, very often between 15 and 20 per cent) but the market 'pie' is not large enough to satisfy all the corporate growth objectives.

Thus the international context seems to be destined to become even more competitive, and only the strongest firms will be able to attain their targets. Acquisitions and agreements will play a major role in the company strategies. Companies choose to grow at the international level to benefit from their particular strengths in new markets and sectors, both to strengthen their existing advantages and to create new ones by tapping resources not available domestically. The search for technological, commercial and manufacturing knowledge complementary[15] to that already existing is necessary for an innovative company without production or distribution channels which allow it to transform innovation into a commercial advantage.[16]

The main features of chemical firms' strategies, previously outlined, show the crucial role played by technological progress and R&D. Also many governments are giving considerable assistance to national research programmes. Everybody is trying to build on strengths in terms of the technological base and the portfolio of research projects, always on a global level. Especially in the most technologically advanced sectors, the high cost of research and development of products will emphasize the relevance of technological strategies as far as the competitive performance of a firm is conceived. Overall, this suggests big changes and restructuring in the medium run in the chemical industry.

Notes

1. The Japanese chemical industry is relatively young and has a leadership position only in the sector of plastic materials.

2. The Italian situation seems to be particularly critical because structural factors, such as the low average size of its firms and its insufficient cash flows (and the prominence of public companies), make a short-term solution of these problems unlikely. The Italian chemical industry runs the risk of being competitive mainly in the less developed markets.

3. According to Pavitt's definition of science-based sectors. For a taxonomy of industries according to sources and types of innovation, see Pavitt (1984).

4. The emergence of biotechnologies and new materials as technological trajectors (that is, as technological opportunities outside companies as opposed to Rosenberg technological imperatives to indicate the technological opportunities typical of a given company) can be connected to the three following coexisting conditions (Nelson and Winter, 1977):

(a) the difficulty of progressing further along the old trajectories of chemistry;
(b) the existence of bridge institutions between pure research and industrial application;
(c) the creation of new scientific opportunities in the field of biology and physics.

5. For a survey of relevant facts about the first stages of the chemical industry, see Ciusa (1965).

6. Different lines of development pursued by the chemical companies of the major industrialized countries are examined in Morandi and Pantini (1982).

7. The factors that determined this decline in growth have been examined extensively in the literature. See Azzolini et al. (1979); Morandi and Pantini (1982); CEEP Report (1986).

8. In the first half of the 1970s inflation concealed the weakness of the chemical industry. In fact while total sales, investments and foreign trade were growing, the prices of chemicals were increasing even more rapidly (*Chemical & Engineering News*, 7 April 1986).

9. In the United Kingdom, investments in chemicals, by contrast with most manufacturing industries, grew at a rate superior to that of sales. However, Japan recorded high investments and research growth rates in all manufacturing sectors. The two events were likely to have a cumulative effect because of the increasingly closer relation between investments in R&D and those in fixed capital, making possible an easier and more efficient introduction of innovations.

10. The restructuring of firms' production processes determined both the introduction of automation into industry and a reduction in employment (especially in bulk chemicals). Accumulation favoured investment aimed at the substitution of factors of production over investment aimed at expansion. Analyses of the structure of investments, in fact, indicate that there was a shift from investments in construction towards investments in machinery and equipment and transportation. This shift

reflected the need to modernize and rationalize productive structures.

11. The expected post-oil shock forecast of complete delocalization of petrochemicals in OPEC countries has been partially fulfilled: in fact the industrialization process has been more difficult than expected, and the oil price drop in the 1980s has hampered the investments in these areas. This error has been due to the petrochemical interpretation of the product life cycle (see Stobaugh, 1982). On the effects of oil price drop see Pearce (1983) and Guimet and Orme (1985).

12. In 1986 the US trade balance stabilized, and in 1987 it grew, thanks to the dollar devaluation.

13. The theory of industrial investment diversification to minimize risks has stemmed from the theory of portfolio choices in uncertain conditions introduced by Tobin to diversify financial investments (see Tobin, 1958). The usefulness of this theory is questionable, since it rules out by definition (in portfolio-theory activities are uncorrelated by assumption) the exploitation of economies of scope existing in related productive activities.

14. According to Ohmae (1985), in a global industry characterized by integrated markets and increasing qualitative similarities, it is fundamental to operate in all main worldwide markets, that is, the triad including the US, Europe and Japan. This is certainly true for chemicals, even if in this field Japan has not yet become an international pole.

15. The concept of complementarity in cooperation strategies was introduced by Richardson (1972).

16. See Teece (1986). As to the importance of international cooperation agreements, there is a wide-ranging literature. See for example OECD (1986) or Jacquemin (1986).

References

Azzolini, R., Dimalta, G. and Pastore, R. (1979) *L'industria chimica tra crisi e programmazione*. Rome: Editori Riuniti.

Bower, J.L. (1986) *When Markets Quake*. Boston, MA: Harvard Business School Press.

Brancadore, V. and Schieppati, G. (1983) 'Le strategie industriali per gli anni '80: il settore chimico di base con particolare riferimento alla petrolchimica', *Economia e politica industriale*, 39: 173–92.

Cainarca, G.C. and Mariotti, S. (1985) 'I sentieri della diversificazione orientata. Un'analisi empirica dei comportamenti delle grandi imprese statunitensi', *Economia e politica industriale*, 46.

Caves, R.E. and Porter, M.E. (1977) 'From entry barriers to mobility barriers: conjectural decisions and contrived deterrence to new competition', *Quarterly Journal of Economics*, 4: 241–61.

CEEP Report, (1986) (Rapporto CEEP) (ed. by R. Gallo) *Industria chimica: strategie, struttura economica e finanziaria*.

Ciusa, W. (1965) *Trattato di merceologia. Aspetti tecnici ed economici dei più importanti cicli chimico-produttivi*. Turin: Utet.

Ergas, H. (1984) 'Corporate strategies in transition', in A. Jacquemin (ed.), *European Industry: Public Policy and Corporate Strategy*. Oxford: Clarendon Press. pp. 327–42.

Federchimica (1985, 1986) 'Rapporto sullo stato dell'industria chimica in Italia'.

Guimet, J. and Orme, P. (1985) *Petrochemical Industry: Energy Aspects of Structural Change*. Paris: OECD.

Jacquemin, A. (1986) 'Compétition européenne et coopération entre entreprises en matière de recherche-développement'. Document of EEC Commission, Brussels.

Kline, C. (1972) 'Maximizing profits in chemicals', *Chemtech*, February: 39–67.

Linn, R.A. (1984) 'Product development in the chemical industry: a description of a maturing business', *Journal of Product Innovation Management*, February: 116–28.

Little, Arthur D. (1981) 'The chemical industry: business strategies for the 1980s'. Cambridge, MA, April.

Little, Arthur D. (1986) 'Pacing developments in the chemical industry to 2000 and beyond'. Cambridge, MA, July.

Mansfield, W. (1986) 'Petrochemicals and chemicals', in H. Peter Gray (ed.), *Research in International Business and Finance: a Research Annual*, vol. 5. Greenwich, NY: JAI Press. pp. 191–221.

Mariotti, S. (1986) 'La diversificazione delle imprese industriali europee e statunitensi: una comparazione', *Economia e politica industriale*, 52: 113–49.

McGee, J. and Thomas, H. (1986) 'Strategic groups theory, research and taxonomy', *Strategic Management Journal*, 7 (2): 141–60.

Mondo Economico (1984, 1987) Monthly reports on chemical industry, 31 May 1984, 27 April 1987.

Morandi, L. and Pantini, G. (1982) *Dialogo sull'industria chimica. Vie e modelli di sviluppo*. Milan: ETAS Libri.

Mussati, G. (1983) 'Strategie d'impresa e processi di internazionalizzazione dopo le crisi petrolifere: il caso dell'industria chimica italiana', *Economia e politica industriale*, 40: 179–99.

Nelson, R.N. and Winter, S.G. (1977) 'In search of a useful theory of innovation', *Research Policy*, 6: 36–76.

OECD (1986) *Technical Cooperation Agreements between Firms: some Initial Data and Analysis*. Paris: Directorate for Science, Technology and Industry.

Ohmae, K. (1985) *Triad Power: the Coming Shape of Global Competition*. New York: Free Press.

Pavitt, K. (1984) 'Sectoral patterns of technical change: towards a taxonomy and a theory', *Research Policy*, 13: 343–73.

Pearce, J. (1983) *The Third Oil Shock. The Effect of Lower Oil Prices*. Royal Institute of International Affairs.

Porter, M.E. (1979) 'The structure within industries and companies' performance', *Review of Economics and Statistics*, 214–27.

Porter, M.E. (ed.) (1986) *Competition in Global Industries*. Boston, MA: Harvard Business School Press.

Ranci, P. and Vaccà, S. (eds) (1979) *L'industria petrolchimica italiana: anatomia di una crisi*. Milan: Franco Angeli.

Richardson, G.B. (1972) 'The organisation of industry', *Economic Journal*, September: 883–96.

SRI International (1985) 'Changing strategies for chemical companies'. Paper presented at the 1985 Industry Consulting Division Symposium, Zurich, Switzerland, November.

Stobaugh, R. (1982) 'Il ciclo di vita del prodotto, i flussi tecnologici internazionali ed il settore petrolchimico', *Quaderni dell'industria chimica*, pp. 15–41.

Teece, D.J. (1986) 'Innovazione tecnologica e successo imprenditoriale', *L'Industria*, 4 (October): 605–43.

Tobin, J. (1958) 'Liquidity preferences as behavior towards risk', *Review of Economic Studies*, February: 65–87.

3

Associational Systems in the Chemical Industry

Wyn Grant

The Organization of Business Interests project was concerned, among other questions, with the internal organizational structures of business interest associations, and the linkages between such associations. The research design for the project noted: 'Business interests in developed capitalist societies are organized in a great number and variety of associations that are linked together in a highly complex pattern of inter-organizational relations' (Schmitter and Streeck, 1981: 140). In order to make sense of this complexity, and the factors underlying it, it is necessary to analyse the organizational properties both of individual chemical industry associations, and of the system of associations as a whole.

Schmitter and Streeck defined (1981: 183) an associational system as 'the universe of BIAs representing interests from a particular sector'. They pointed out that associational systems may be more or less differentiated or complex. 'While differentiation produces *complexity*, integration *orders* that complexity and binds different units together in *coordinated patterns of interaction*' (Schmitter and Streeck, 1981: 183). In this chapter there is therefore an examination of whether integration of production in chemicals has influenced relations between associations, and of whether pressures towards differentiation differ within the sector across countries.

There is not space in this chapter, nor would it be particularly fruitful, to examine all the organizational properties of the individual associations. Indeed, not all the data collected in the project yield interesting or useful results. Rather, attention is focused on resources as key determinants of the ability of associations to handle complexity, leading on to an examination of the characteristics of the outputs of chemical industry associations. The chapter concludes with an examination of industry organization at the EC level, which is of particular importance in an international industry such as chemicals.

The particular characteristics of the systems of business interest associations in industrial chemicals can best be appreciated in comparison with those to be found in a very different industry: construction. Industrial chemicals is a capital-intensive industry; construction is labour-intensive. Industrial chemicals is highly internationalized in terms of patterns of ownership and trade; construction is characterized by domestic ownership and is still significantly oriented towards domestic markets. Research and development is of considerable importance in the industrial chemicals industry; construction still relies heavily on traditional techniques.

One of the characteristics of the industrial chemical industry is that the production process is highly integrated. The series of transformations which occur in the processing of basic feedstocks (such as oil or natural gas) into bulk polymers (such as the polyethylenes) often take place on one site in a series of linked plants, or at least on a series of sites linked by pipelines, such as the network of pipelines to be found running between plants in the Rhine–Ruhr area and the Netherlands–Belgian Delta zone centred on Rotterdam. Raw materials are, of course, also conveyed by boat, lorry and rail, but the basic picture is one of a complex interchange of both raw materials and intermediates:

> the Dow establishment in Rotterdam is supplied with raw materials by Dow-Terneuzen; the fertiliser factories of BASF in Antwerp obtain ammonia from the Ammoniak Unie at Pernis (in which BASF participates); Shell-Pernis is the supplier of raw materials to the isocyanate factory of Bayer and Shell in Antwerp; Oxirane in Rotterdam delivers propylene oxide to Bayer in Antwerp, Solvay provides Dow-Terneuzen with chlorine and so on. (Molle and Wever, 1984: 131)

In contrast, production in construction is segmented among a series of trades with their own particular skills and often among competing technologies and skills within a particular trade (for example, different forms of roofing). It is therefore not surprising to find that there are far fewer first-order associations in industrial chemicals than in construction: the median number of such associations in the countries studied is fifteen in chemicals and sixty-seven in construction. In terms of relationships *between* associations, the greater integration of production in chemicals does not seem to make very much difference in countries which have generally coherent systems of business associations such as West Germany. It would appear that there is some indirect influence in countries which have less coherent associative systems such as the United Kingdom. In that country only 50 per cent of first-order associations are affiliated to a higher-order association in construction, but this figure increases to 75 per cent in industrial chemicals.

The nature of the production process in chemicals leads to the geographical concentration of production in a relatively small number of locations; construction, by its nature, is a geographically dispersed activity. It has been noted that:

> The Ruhr area and the Rotterdam–Antwerp delta in 1980 accounted for more than 35 per cent of total West European ethylene capacity; if the units at Beek, Frankfurt-Ludwigshafen and in Northern France . . . are included, the percentage goes up to 50. The triangle Rotterdam–Ludwigshafen–Le Havre can indeed with some justification be called the Western European petrochemical triangle or quadrangle. (Molle and Wever, 1984: 90)

It is therefore not surprising that there is a considerable number of territorially differentiated associations in construction, whereas the associations are much less common in chemicals. Four of the nine countries studied have no such associations in industrial chemicals, and the one country with a large number of such associations (Italy with thirty-one) has a strong territorial emphasis in the general organization of its business associations. In construction, four countries have more territorially differentiated associations than the highest score in chemicals. Five of the nine countries have more than 50 per cent of their associations in construction differentiated by territory; in industrial chemicals only Italy has a score above 50 per cent.

All countries have at least two associations specialized by product in industrial chemicals, with a median score of 11. However, reflecting the dominant position of large firms in the industry, and their ability to cross-subsidize the provision of selective incentives to smaller firms, only three of the nine countries have associations specialized by firm size in industrial chemicals. In construction, by way of contrast, only two countries do *not* have associations specialized by firm size. Of the countries studied, Canada is the only one that does not have either an employers' association or a combined employers'/trade association in chemicals; all the other countries have at least one association involved in industrial relations.

There is far more competition for members between associations in construction than in chemicals, whether measured in terms of the number of competitive spaces or the number of associations in competition. Four of the nine countries have three or less associations in competition in industrial chemicals. The only really high score is in Italy with thirty-three associations in competition, but even here there is less competition than in construction.

The greater homogeneity of the associational system in chemicals is confirmed by the construction of an index of overall differentiation.[1] The range of scores in industrial chemicals (66.67–123.53) is both relatively low and narrow, suggesting that the pressures tending towards more differentiation do not differ greatly from one country to another. In construction the scores are both higher and more differentiated by country (97.22–293.73).

A higher proportion of associations in chemicals is made up of pure and mixed higher-order associations than in construction. The median score in construction is 10.52 per cent; in industrial chemicals it is 32 per cent. This would suggest that in industrial chemicals there is greater capacity for the effective articulation of views across the industry as a whole than in construction. Moreover, as will be noted in a later section, the industry is effectively organized at a European level.

One important qualification must be made to this general picture of relatively coherent and well-organized associational systems in industrial chemicals in most countries, with national differences less pronounced than in a sector such as construction. Our study concentrates on industrial chemicals (data were also collected on pharmaceuticals and agrochemicals). Industrial chemicals is the core of the chemical industry and its best-organized subsector, reflecting the integrated nature of the production process and the dominance of a small number of well-resourced companies. It is not surprising that it is able to create a relatively well-integrated associational system in most countries.

When one moves away from the core of the industry to such subsectors as paints and soap and detergents a different pattern emerges. First, these subsectors are often customers of the industrial chemicals industry; they use raw materials produced by it for the manufacture of their own products. Their interests are often at variance with those of industrial chemical companies: for example, a rise in the price of essential raw materials can adversely affect their profitability. Second, although the large chemical companies are present in these subsectors, they are much less dominant and able to exert less influence over the shaping of associative arrangements. Given these factors there is greater variety in the organization of these peripheral subsectors; in particular, national differences become more pronounced. For example, although there is not great variation in the associational representation of industrial chemicals in Britain and West Germany, sectors such as paints, and soaps and detergents, have a much closer relationship with VCI than with its British counterpart.

Input and output characteristics of chemical industry associations

As well as considering the characteristics of associational systems in the chemicals industry, in terms of the distribution of responsibilities between associations and the relationship between them, it is also important to consider the salient characteristics of individual associations from an aggregate perspective. In particular it is important to consider the resources available to individual associations and the outputs they produce from those resources. As far as outputs are concerned, attention will be focused on those activities which indicate a capacity on the part of associations to act as effective intermediaries between their members and governments.

Resources

Resources will be considered in relation to the incomes of associations and the size and skills of their staffs. Clearly the two are interrelated: an association with a very small income may not be able to afford any full-time staff, and may depend on another association or a firm for office services; an association with a large income will be in a position to employ a specialized staff, with considerable differentiation of responsibilities within the organization. However, one must be wary of assuming that a large staff can necessarily be equated with effectiveness in terms of defending the interests of the organization's members. Large professional staffs may develop their own bureaucratic goals which affect the way in which they discharge their responsibilities; expansion of the organization's tasks may displace other goals which are more relevant to members' interests. Moreover, time may be spent on communications within the association bureaucracy at the expense of communications with outside environments which the association is trying to influence.

As one would expect, there is considerable variation in the incomes of associations in industrial chemicals, ranging in 1980 from $12,000 a year to $17 million (the VCI in West Germany). However, the distribution of incomes among associations seems to be very little different from construction: in both industries just under 30 per cent of associations had an income in 1980 of over $1 million a year. In rank-order terms the German VCI is followed by Britain's CIA, which had just over a third of the VCI's income in 1980; it is followed by the leading Italian association with just under half of the CIA's income. Another German association takes fourth place, followed by the Swedish peak association.

Two-thirds of all associations have less than ten staff, but there are three with over 100 staff (the VCI, the CIA and Switzerland's SSIC/SGCI). The distribution of staff numbers between associations is very similar to that in construction. However, somewhat more construction associations have no graduates on their staff (a fifth as against a tenth in industrial chemicals). Another difference between the two industries is in terms of the high turnover of members of construction associations, reflecting both the larger number of small firms in that industry (smaller enterprises are more likely to join and leave associations on the basis of a rationalistic calculation of costs and benefits), and the higher bankruptcy rate in construction. Over 80 per cent of the industrial chemicals associations had an exit ratio of less than 3 per cent, compared with 58 per cent of construction associations. Ten per cent of all construction associations had an annual exit ratio of over 10 per cent, compared with only one (Swiss) chemicals association.

Collective goods

Like most business interest associations the chemical industry associations studied were heavily involved in making representations to executive bodies (95 per cent of all associations) and, to a somewhat lesser extent, to legislatures (70 per cent of all associations). However, the data also reveal some interesting differences between chemical industry associations and construction industry associations in terms of representational activities which benefit all firms in an industry regardless of whether or not they are association members.

Forty-six per cent of all chemical industry associations were involved in the provision of information to government bodies, compared with only 16 per cent of construction associations. This suggests that government is more dependent on information provided by business associations in chemicals than it is in other industries. Such a finding is not surprising when one considers the highly technical character of many of the problems that arise in relation to the industry. One implication is that chemical associations may be more intimately bound up with the decision-making process within government, and hence better able to exert influence on behalf of their members.

The scientific character of the industry is reflected in the fact that 78 per cent of the associations studied make representations to scientific agencies. For 38 per cent of associations this was a major activity, compared with only 16 per cent in construction. The international character of the chemical industry was evident in the fact that over half of the associations sampled made representations to supranational institutions, whereas in construction 80 per cent had no such involvement. Moreover, representation to supranational institutions was a major activity for over a quarter (27 per cent) of chemical associations, but for none in construction.

The sensitivity of the chemical industry to its public image, and concern about what has been termed 'chemophobia' among the general population, is reflected in the fact that 70 per cent of associations were involved in public relations activities on behalf of the industry, with this being a major activity for 41 per cent of associations, twice the level in construction. However, chemical associations were *less* likely to be involved in representations to customers and suppliers than construction associations. Seventy-eight per cent of chemical associations were not involved in presentations towards suppliers, which is not surprising when one considers that the industry generates a considerable proportion of its own raw materials (the major exception is energy, which is often provided by state-owned entities). Similarly, 62 per cent of chemical associations were not involved in representations towards customers, reflecting the downward integration of the industry.

It was estimated by the research teams that 79 per cent of chemical associations expended over 50 per cent of their organizational effort on the provision of collective goods, whilst in construction 73 per cent of associations were estimated to expend *under* 50 per cent of their organizational effort on collective goods provision. Just over a third of chemical associations expended over 75 per cent of their organizational effort on collective goods provision, compared with just over 10 per cent of construction associations. This emphasis on collective goods provision is not surprising when one considers that the large firms which dominate the chemical industry are less likely to be attracted to association membership by the provision of selective incentives than the many smaller firms in construction. For a large chemical firm, even quite a large association subscription is likely to amount to a very small proportion of turnover. Concern about the cost of paying for benefits which would be available without membership is likely to be outweighed by a belief in the desirability of effective representation to government, possibly reinforced by close contacts among key decision-makers in the leading chemical companies.

Selective goods

Selective goods provision (that is the provision of services restricted to members and/or sold to non-members) was, therefore, less important to chemicals associations than to associations in industries such as construction. The selective good most frequently provided was general economic information on the state of the industry (62 per cent of associations studied). Technical and economic consulting services were provided by 60 per cent of associations, but tended to rank as a minor service. The third most important selective good provided by associations was in the area of research (41 per cent of associations). Advice on legal, tax and insurance matters was provided by 35 per cent of all associations sampled, a lower figure than in construction, where nearly half of all associations provided such services, which are particularly likely to appeal to smaller firms. Only one chemicals association expended more than 50 per cent of its organizational effort on the provision of selective goods, compared with nearly a quarter of all construction industry associations. Indeed, 72 per cent of chemical associations for which estimates were available expended 25 per cent or less of their organizational effort on the provision of selective goods. With the exception of the exchange of information on prevalent economic conditions in the industry, selective goods provision is clearly not a major area of chemical association activity.

Monopoly goods

Monopoly goods are defined in the project research design in terms of the provision of

> private selective goods through the *assistance of other organizations, state. . . . As far as the state is concerned, such support consists primarily in monopoly rights* on the provision of certain vital goods or services to (potential) BIA members, or a share in the *authority* of the state to make legally binding decisions on matters of interest to BIA constituents. (Schmitter and Streeck, 1981: 229)

The provision of monopoly goods by associations is of particular importance in the context of the business interests project as it indicates a high capability on the part of associations to act as effective intermediaries between their members and government. In particular, the provision of monopoly goods suggests an ability either to undertake delegated public policy functions, or to perform functions which might otherwise have to be discharged by the state. Schmitter and Streeck note:

> If BIAs want to become stable and effective interest organizations, it is essential that they acquire the capacity to procure the compliance of their members with negotiated agreements on matters which otherwise would be entirely under their discretion as private owners. (Schmitter and Streeck, 1981: 236)

The research teams were asked to estimate the share of

organizational effort spent by each of the associations in their country on the provision of monopoly goods. Such a measure is obviously open to variations in its application, but if systematic differences between sectors are discovered, they merit further exploration. In fact there were significant differences between chemicals and construction. Over half of all industrial chemicals associations (52 per cent) provided no monopoly goods at all, compared with only just over a quarter in construction (26.5 per cent). The highest level of monopoly goods provision by any one association in chemicals was 40 per cent, whereas 11 per cent of construction associations expended over 50 per cent of their organizational effort on the provision of monopoly goods.

It would seem that the construction industry has particular characteristics which ease the acquisition and development of private interest government functions by associations. These differences in comparison to the industrial chemicals industry are particularly apparent in four areas of monopoly goods provision. Over one-third (35 per cent) of construction associations are involved in monopoly goods provision in the area of wage agreements, compared with only 14 per cent of chemicals associations. This difference reflects the greater importance of small and medium-sized firms in construction, which are more likely to band together to negotiate on a sector-wide basis, rather than allow themselves to be picked off one by one by the unions. In chemicals, many large firms such as ICI prefer to conduct their own wage negotiations and not to participate in binding industry-wide agreements.

A second striking contrast in monopoly goods provision is to be found in the area of quality control. Nearly 30 per cent of construction associations are involved in the enforcement of quality standards, compared with 11 per cent of industrial chemicals associations (two of those undertaking such a task as a major activity are to be found in Sweden). This is not surprising when one considers that a central concern of many construction associations is the suppression of 'cowboy' firms which undercut their members and give the industry an unfavourable image through bad workmanship. In the chemical industry customers will soon complain if products do not meet the specified standards, as the results will quickly be apparent in the production process. Of course, in the pharmaceutical branch of the chemical industry, which is not considered here, there is likely to be more associational involvement in quality control in partnership with the state (see de Vroom, 1985). A somewhat related area is certification and licensing, which involves 22 per cent of construction associations, but only one industrial chemicals association.

A third area of contrast is vocational training, which involves 20 per cent of construction associations in the provision of monopoly goods, but only 5 per cent of industrial chemicals associations. This contrast reflects the character of the construction industry which is skill-intensive with a highly mobile workforce; hence firms have an interest in the collective provision of training. In industrial chemicals, major firms in the industry are more likely to meet their own training needs.

Perhaps the most striking difference is to be found in the area of control of competition, where 37 per cent of construction associations are involved in the provision of monopoly goods, compared with 8 per cent of industrial chemicals associations. This difference may be a little misleading in that governments may have been more willing to legitimize cartel-like arrangements in construction, given the disruptive consequences of unrestrained competition in that industry, whereas control of competition in industrial chemicals may have taken more illegal or, at any rate, less tolerated forms. Even if that is the case, however, the consequences for associational development are significant because, in construction, associations are able to be openly and formally involved in the control of competition, whereas such an involvement is much less common in chemicals.

Having drawn attention to these differences, it should be pointed out that there are a number of areas where associations are not generally involved in monopoly goods provision in either industry. In particular this is true of investment control, an area which lies at the heartland of the commercial autonomy of the firm.

In terms of the delegation of tasks by the state to associations (what Schmitter and Streeck call 'authoritative goods'), 30 per cent of construction associations enjoy such delegated authority in the area of vocational training, compared with only 13 per cent of industrial chemicals associations. Moreover, unlike construction, none of the chemicals associations are able to discriminate between members and non-members. Interestingly, in the area of quality control there is relatively little delegation by the state, suggesting that this is an area where associations have developed on their own initiative a capability to make decisions which are binding on their members.

One interesting finding is that twice as many industrial chemicals associations as construction associations (16 per cent compared with 7 per cent) are involved in delegated tasks in the area of industrial policy. This suggests that associations in the chemical industry are able to make some contribution to the important task of restructuring and rationalization. Moreover, much of the activity in this area is likely to be undertaken by European-level associations which were not part of the database. (For further discussion of the role of European-level associations in the restructuring process, see Chapter 10, by Grant, on the overcapacity problem in the petrochemicals industry.)

Even allowing for some involvement in industrial policy questions, the general picture that emerges is of associations that are well-equipped to undertake representational tasks on behalf of the industry, but do not exert much authoritative control over their members. This is not surprising when one considers that, unlike the construction industry, the industrial chemicals industry is dominated by large, international firms which are not likely to be willing to surrender much of their autonomy of action to associations. Questions such as sectoral adjustment are more likely to be dealt with on the basis of intra-company rationalization and inter-company deals. The contribution of associations to this process, especially at a European level, has largely been in the area of information exchange (within the limits permitted by cartel law) and lobbying efforts to ensure that the rationalization process is facilitated by the EEC (see Chapter 10, by Grant, on the over-capacity question). Associative activity has indirectly contributed to the rationalization process, but its role should not be exaggerated.

Indeed, an executive of a major firm argued in interview that a preliminary paper from the project had been too generous in its assessment of a national association, arguing that the importance of the association was at a working level in relation to issues such as hazard warnings on road vehicles. The firm was very used to dealing direct with national government departments, although at a European level it relied more on the Conseil Européen des Fédérations de l'Industrie Chimique (CEFIC). This body, it was argued, had more purchase on the Commission than the national association had on the country's government. It is to the role of such international organizations that I shall now turn.

Industry organization at the European level

In terms of patterns of trade the role of multinational companies, and the way in which markets are perceived and organized, industrial chemicals is one of the most internationalized of all industries. Supranational forms of industry organization therefore acquire a special significance. However, the three main production areas in the world (Western Europe, the United States and Japan) have somewhat different interests: for example, Western Europe exports a far larger proportion of its output than the United States. American firms, on the other hand, are much more extensively involved in joint ventures with the new producers in the Middle East, particularly Saudi Arabia. These new petrochemicals producers represent a long-run challenge to the West European industry's export effort. Hence, although there are frequent contacts and exchanges of views between the industry associations in Europe, the United States and Japan, particularly on environmental and regulatory issues, where they have more interests in common, this does not take place within any global organization. However, the International Plastics Association Director (IPAD) conference does provide an annual forum in which directors of plastics associations from all over the world can exchange views.

The principal representative body for the chemical industry at the European level is a federation of fifteen national associations, CEFIC. Before 1972 there were separate associations for EEC and EFTA countries; the combined organization has generally worked well, although sometimes the non-EEC members think that it is too oriented towards EEC questions. CEFIC has more than fifty sector groups, some of which are organized as associations (for example, the European Catalysts Manufacturers' Associations) and some are not (such as the Ethylene Oxide and Derivatives Sector Group). By far the largest of these sector groups is the Association of Petrochemicals Producers in Europe (APPE), which is discussed further below. It should be noted that other parts of the chemical industry, such as paints, agrochemicals, soaps and detergents, have their own European-level organizations. However, the industrial chemicals industry, which is our concern in this book, is covered by CEFIC and its affiliates.

CEFIC's offices are in Brussels, where in 1985 it had a staff of forty-two, making it one of the better-resourced industry associations operating at the European level. However, it has not been without its internal organizational problems. In particular these arise from having a federation of federations represent an industry which is dominated by a relatively small number of large international firms. There has been a feeling in the past that the officials of national organizations have had too great an influence on the development of CEFIC policy compared with the representatives of the major companies. This issue was tackled by a number of internal organizational changes made under the presidency of Sir John Harvey-Jones of ICI (1984–86), which gave individual companies an enhanced role in the decision-making process.

CEFIC is incorporated under Belgian law as a non-profit-making organization with scientific aims, and formal power resides in its

General Assembly. In practice, however, the most important issues were dealt with by a Committee of Directors (of national associations) and a General Policy Committee grouping senior company executives. It was felt that this arrangement gave too much weight to the association directors. Hence, in 1985 CEFIC bye-laws were changed to create a new category of corporate associative members (companies operating on a transnational and European scale). An Assembly of Corporate Associate Members elects its own representatives to take part in an advisory capacity in the meetings of the Committee of Directors.

Another important change which recognized the importance of major companies in the industry was the formation of APPE in February 1985. APPE organizes the leading petrochemicals producers in Europe and has its own director-general operating from the CEFIC offices. This organization brought together several formal and informal committees which previously existed within the CEFIC structure. It provides a mechanism for achieving a coherent representation of the particular interests of the petrochemical industry on such matters as overcapacity problems and relations with the new producers in the Middle East and elsewhere.

CEFIC's representational efforts are, of course, primarily directed towards the EEC, and in particular towards the Commission. DG III has responsibility for the chemical industry, but contacts are maintained with a number of other directorates; for example, DG I on external trade relations, DG VI on the impact of the Common Agricultural Policy on the chemicals industry, DG XI on environmental questions, and DG XVII on energy issues. CEFIC is also in regular contact with a number of other international organizations; for example UN bodies such as IMO on transport safety issues, and GATT on trade questions.

Divergences of national style in the approach to industrial adjustment questions naturally affect discussion within CEFIC. For example, the Dutch tend to adopt a liberal approach, whereas the French have more protectionist sentiments. State owned companies have to pay greater attention to considerations of national interest than privately owned transnationals. Nevertheless, the industry's operating environment is considerably affected by decisions taken by the EEC and other international bodies, and it was clear from interviews with companies that CEFIC was of considerable importance to them; perhaps of even greater importance than the national federations. However, it will be a very long time before, as one informant suggested, CEFIC becomes an integrated body with the national chemical federations as subordinate components. Political organization tends to lag behind the realities of economic organization in an internationalized industry.

Note

1. The index of overall differentiation was constructed from a number of variables used in the IIM project. The numbers of associations specialized on the following dimensions were aggregated: territory, product, firm, size, political allegiance, religion, other criteria. The total obtained was multiplied by 100, and then divided by the number of membership associations in the sector.

References

Molle, W. and Wever, E. (1984) *Oil Refineries and Petrochemical Industries in Western Europe*. Aldershot: Gower.

Schmitter, P.C. and Streeck, W. (1981) 'The organization of business interests: a research design to study the associative action of business in the advanced industrial societies of Western Europe'. Berlin: International Institute of Management labour market policy discussion papers.

de Vroom, B. (1985) 'Quality regulation in the Dutch pharmaceutical industry: conditions for private regulation by business interest associations', in P.C. Schmitter and W. Streeck (eds), *Private Interest Government*. London: Sage. pp. 128–49.

4

Political Turbulence, Enterprise Crisis and Industrial Recovery: ICI and Montedison

Wyn Grant and Alberto Martinelli

ICI and Montedison are both major international chemical companies, but there are some significant differences in their histories, operating performances and political experiences. ICI was formed from a merger of four companies in 1926 to provide a counterpart to IG Farben of Germany. Montedison was formed in 1966 through a merger between Montecatini and Edison. Montecatini had developed extensive chemical interests since its formation in 1888, and had built Italy's first petrochemical complexes in 1949. Edison was originally an electricity utility, but had diversified in anticipation of nationalization, which took place in 1962.

Both ICI and Montedison experienced major restructuring in the early 1980s, retreating from loss-making commodity operations and increasing their presence in more profitable specialty chemicals. For example, ICI made sixty-two acquisitions and fifteen divestments between 1982 and 1986 in pursuit of its aim of reshaping the group. ICI had enjoyed a record of considerable, if not unblemished, success up to 1980 when the company incurred losses for the first time in its history. Montedison was in much worse shape at the beginning of the 1980s, having suffered from the 'chemical war' between Italian companies and from an over-politicized operating environment. ICI faced problems resulting from the rapid appreciation of the sterling exchange rate between 1978 and 1981, coinciding with a sharp rise in labour costs and the onset of the world recession. In the UK ICI estimates that it 'lost about 20 per cent of our customer base in our petrochemicals and general chemicals business' (Harvey-Jones, 1985: para. 5). Against this difficult background, both companies made substantial progress in reorienting their commercial strategies, and tackling their basic problems.

The two companies compared

Product profiles

Both companies have made a considerable effort to shift away from a dependence on commodity areas of activity. Basic chemicals accounted for 42 per cent of Montedison's sales in 1982; by 1984 their share was estimated at 36 per cent; and a further fall to 30 per cent was planned for by 1987. Montedison has placed considerable emphasis on developing its technological strengths in specialty chemicals and high-performance materials, an area in which there was a 22 per cent increase in sales in 1985 compared with 1984. Similarly, ICI has reduced its presence in commodity chemicals and strengthened its specialty chemicals operations. Industrial chemicals accounted for 49 per cent of ICI's sales in 1982, compared with 22 per cent of sales in pharmaceuticals and other consumer and specialty products. By 1985 the balance had changed to 44 per cent for industrial chemicals and 28 per cent for consumer and specialty products. Moreover, the latter category accounted for 42 per cent of profits. Even so, in 1987 ICI set up a new chemical and polymers group, scheduled to become a limited company in its own right in 1988, to bring together the non-specialist, bulk products areas of the company. The reorganization was intended to mark a clear separation between the traditional, commodity chemicals side of ICI and its activities in newer, faster-developing areas such as pharmaceuticals, composite materials and electronics.

However, both companies still have a considerable level of involvement in petrochemicals and plastics, although Montedison hopes to dispose more of its activities in this area to Enichem, whilst ICI has set up a joint venture with that company as a part of an effort to rationalize PVC production. ICI is more strongly involved in agrochemicals than Montedison, whereas the Italian company has a stronger involvement in energy activities. One significant difference is that 16 per cent of Montedison's sales in 1985 came from 'services', including real estate; ICI has a policy of avoiding diversification into areas not related to its strengths.

Employment and sales profile

Both companies have been reducing numbers employed as part of their programmes of restructuring and reorientation. Montedison had 69,653 employees in 1985, compared with 86,574 in 1981. ICI has made similar reductions in its workforce: down from 132,400 in 1981 to 118,600 in 1985. It should be noted that less than half of ICI's employees are in the UK.

This international orientation of ICI is also reflected in its sales profile: 25 per cent of ICI's chemicals customers in 1985 were located in the UK, with 26 per cent in North and South America and 22 per cent in Continental Europe. By contrast, 60 per cent of Montedison's revenue in 1984 came from Italy, with 21 per cent from the rest of Western Europe, and 5 per cent from North America. However, it is interesting that Montedison's joint venture with the American company, Hercules, called Himont, is setting up marketing companies in the UK, France, Germany and Hong Kong. Although Montedison is not well placed for sales to North European markets, it is increasing its presence in the North American market (note its acquisition of Conserv, a Florida fertilizers company). Both ICI and Montedison are clearly major players in the international chemical industry.

The bottom line: financing and profitability

The strongest contrast between the two companies is to be found in their financial arrangements and profitability. After taking account of interest payments, Montedison made a (reducing) loss in 1982, 1983 and 1984, making a modest profit in 1985. The year 1982 was a particularly difficult one for ICI, but even then the company made a profit after tax and extraordinary items. ICI's position improved in subsequent years, with pre-tax profits of over £1000 million being recorded for the first time in 1984. Even if one disregards Montedison's problems arising from interest payments, there is a clear difference in operating profit margins between the two companies. ICI's margins in 1984 were over twice those of Montedison in 1984 (14.9 per cent against 6.2 per cent) and almost twice as high in 1985 (12.9 per cent against 6.5 per cent).

In 1986 Montedison still had to restore the health of its balance sheet. The debt/equity ratio had fallen from 2.63 in 1982 to 1.92 in 1985, and short-term debt as a percentage of total debt had fallen from 42 per cent to 11 per cent over the same period. However, whilst acknowledging that 'management is beginning to get on top of the problem', a respected industry commentary noted that 'there is still a long way to go – even the achievement of the 1987 forecast of a 1:1 debt/equity ratio would leave debt still at a high level' (W. Greenwell & Co., Company Comment on Montedison, 10 April 1986). ICI, by way of contrast, has considerable financial autonomy. Company policy is to fund over 80 per cent of investment from reserves (Harvey-Jones, 1985: Q. 1299). The company has maintained substantial reserves which saw it through the period of difficulty at the beginning of the 1980s. Perhaps the main financial constraint it faces is the existence of an active equity market in the UK which is very sensitive to changes in company performance and, in particular, dividend levels.

Montedison's development as a company

In recent years Montedison has provided a remarkable example of business recovery, changing significantly its own strategy and structure. Montedison is a diversified corporation with activities distributed in several sectors. We will, however, focus on chemicals, which are Montedison's main activity. The chemical industry is itself a mix of businesses very different in strategic terms, and the way in which Montedison has moved in this area is an indicator of its strategic capabilities.

In order to understand the reasons for this recovery one has to focus both on internal changes and on the transformation of the relevant environment (domestic and international, economic and sociopolitical). The evolution and supply and demand in the chemical sector, the competitive system and the overcapacity issue are dealt with in other chapters of this volume. We will focus only on those issues which are necessary in order to understand the strategic choices made by Montedison.

Unlike ICI, which has a rather settled history of development, without erratic political interferences, Montedison had a much more troubled history, due to a complex merger, a more politicized environment and the industrial problems typical of a late developer. The analysis of Montedison's change of strategy in the 1980s thus requires a more detailed reconstruction of what happened in previous decades.

We will identify four major phases in Montedison's history:

1 The phase prior to the merger.
2 The period going from the formation of Montedison by a merger between Montecatini and the Edison company to the oil crisis, a period characterized by the 'oligopolistic war' in Italian chemicals.
3 The first attempts to correct the course of action of Montedison, and the stage of 'rethinking the strategy' between 1975 and 1980.
4 The basic changes introduced under the leadership of Schimberni, from 1980 to the present.

The oligopoly war in Italian chemicals and the oil crisis

The oil crisis of 1974 created serious problems for all chemical companies, all over the world, but Montedison was more vulnerable to these problems for three main reasons: first, because of the oligopoly war which had taken place in Italy in an earlier period; second, because of the peculiar connection between political and economic power; and third, because the recent merger had thus far prevented the formation of a homogeneous management and a unified corporate culture. By way of contrast, ICI's problem was rather different. It *did* have a unified culture, but it was not one conducive to change. ICI's managers had grown up in an era of rapid expansion of the chemical industry; they had no experience of handling contraction. Moreover, the dominant management culture of the company emphasized technological prowess at the cost of marketing (see Pettigrew, 1985).

The two original business groups of Montedison were structurally and culturally very different. Montecatini was a large chemical and mining company, established in 1888, which had diversified from mining into fertilizers, paints, pharmaceuticals, fibres and coal-based chemicals in the period between the two world wars, both through internal growth and acquisition. During the Fascist autarchic regime, and with the formation of a unified protected domestic market, Montecatini reached monopoly positions in many chemical sectors, but at the time developed a structure fragmented into hundreds of small plants scattered over the whole of Italy.

After the Second World War Montecatini built the first petrochemical complexes in Italy. In 1952 Giulio Natta, one of the company's chemists, discovered a way to produce a superior form of polypropylene, for which he later won a Nobel prize. With this strong technology, and its strong competitive position in the domestic market, Montecatini expanded rapidly as plastics were developed worldwide, and Italy developed a booming economy in the 1950s and early 1960s. But, in the same period, Montecatini's monopoly position was attacked by the entry of Edison and ENI in the chemical sector. In some cases, as in fertilizers, the conflict was severe, with resulting excess capacity, falling prices and financial crises.

With the creation of the European Economic Community the competitive game became international, both in terms of the opportunities presented by much larger markets, and in terms of the threats posed by US and European chemical firms. Montecatini was not equipped to compete on a much larger market, which required a different organization of production and distribution capable of reducing costs, because of economies of scale and experience curves. It reacted by building a large project at Brindisi, but a financial crisis soon followed, leading to a need to find a partner in the mid-1960s.

Edison appeared to be the ideal partner, having substantial financial resources derived from the compensation received from the nationalization of the electricity industry, but without any clear strategy for the use of those resources. Edison, from its foundation in 1886, up until the 1950s, had been almost exclusively an electric utility. In the inter-war period, and in the 1950s, Edison consolidated its role as the leading electric company in Italy, becoming the core of the financial–electric group based in Milan. This group held a hegemonic position in Italian industry, by virtue of its financial power, and its discretionary ability to fix energy costs. Edison exerted a strong influence in Assolombarda, the territorial employers' association of Milan, and through it in Confindustria, the peak association (Martinelli et al., 1981: 242 onwards).

Fearing nationalization of the electrical utility industry, Edison diversified its business in the 1950s, entering petrochemicals (with poor results due to Montecatini's strong reaction), metals, fibres, glass and retailing. Edison's utility business was nationalized in 1962, leaving it with strong liquidity, and a diversified business devoid of its major component.

The merger appeared to be a brilliant solution to the problems of both companies, capable of marrying the technological and productive strength of Montecatini with the financial strength of Edison at a time when huge investments and productive restructuring were needed in Italian chemicals. Each company brought in assets of about $800 million. The new company held a dominant share of the Italian market for fertilizers, plastics, synthetic fibres, several chemicals and metals. However, the new corporation was a conglomerate pursuing a range of diversified activities, with hundreds of scattered plants and firms, each with different legal statuses, corporate cultures and organizational structures. In this situation Montecatini–Edison needed a strong centralized leadership, in order to envisage and implement the new strategy, overcome internal strife and build a unified culture. On the contrary, there was confusion and a struggle for control at the top between management groups coming from two partners.

The period which goes from the merger to the first oil shock is characterized by a relatively unsuccessful attempt to work out a coherent strategy and structure for Montecatini–Edison (Montedison after 1972). These attempts were frustrated by three major sets of factors: first, by the difficulties stemming from the merger; second, by features of the Italian chemical industry; and third, by changes in the relevant social and political environment, most of all the wave of labour struggles of the 'hot Autumn' and the connection between parties and business interests.

The early years of Montedison were marked by a severe struggle between oligopolies in the petrochemical industry. The economic boom had led to rapid expansion of capacity in petrochemicals. The hurried pace of investment, aided by substantial government funding mostly directed to backward regions, favoured the expansion of established firms, and the emergence of new firms mostly in primary chemicals. Fifteen separate petrochemical centres sprang up in various parts of the country, most of them small, badly balanced and insufficiently integrated with each other. In the late 1960s two other large national competitors developed, Sir and Liquigas (later Liquichimica), which by locating their investments in the Mezzogiorno, and by favourable political connections, were able to implement an aggressive policy of expansion vis-à-vis the two traditional oligopolistic firms, the private Montedison and the state-controlled ENI.

This oligopoly struggle took place in a context characterized by labour militancy, which brought wage increases three times larger than those negotiated in earlier periods, a nationwide indexing of wages (the *Scala Mobile*) and tighter constraints on managerial discretionary power through a labour law (the *Statuto dei lavoratori*). The oligopoly conflict, and the rapid rise of labour costs, squeezed profit margins. In Montedison, and in other companies, mild profits turned into losses, leading to declining investments and mounting debts (Montedison's debt/equity ratio rose from 0.4 in 1969 to 1.8 in 1972).

In addition, major chemical companies became the site of complex political exchanges, combining oligopoly competition with political conflict among parties and party factions. The relationship between a rather homogeneous organized business class, and a unified government coalition, which had existed in the early 1950s, had long since given rise to a fragmented network of influences in which different party factions related to different centres of economic power (see Martinelli, 1979). Montedison's president at the time, Eugenio Cefis, was a typical representative of the 'politicized manager'. While envisaging an ambitious plan for the chemical industry, and while introducing some actual changes in Montedison, he also tried to play a 'king-maker' role in the presidential elections of 1971, as well as trying unsuccessfully to compete with Gianni Agnelli, the president of Fiat, for the leadership of Italian organized business.

These far-reaching ambitions made it difficult for him to concentrate on the more specific problem of Montedison's restructuring and rationalization. The rationalization was also made difficult by the third set of factors we mentioned, the difficult coexistence of two managerial groups, and the internal conflicts within Montedison. The combined effect of these factors was that of losing valuable time, exhausting cash flows stemming from the nationalization of the Edison electric utility and increasing confusion at the top of the enterprise.

The oil shock and Montedison's change of leadership

On the eve of the oil shock Montedison had too diversified a spectrum of activities, and too many products with few strengths; a too heavy dependence on basic chemicals; a product structure fragmented in many plants, some of which were obsolete; a low degree of internationalization; an overcentralized organizational structure; a rapidly declining self-financing capability; and an obsolete managerial culture in many sectors of the firm. To give one example of how serious the situation was, a 1972 check-up of crisis points found that eighty-four productive units were not competitive, and that 20 per cent of Montedison's labour force was surplus to requirements. In order to improve the situation Montedison started to simplify its product structure through sales of peripheral activities, mergers and organizational changes. In addition, in 1973, a very ambitious plan of new investments was decided, in order to increase plant capacity. This was, however, later substantially reduced and revised because of the impact of the oil crisis on the Italian economy, and on Montedison's financial structure.

The stagflation stemming from the oil crisis had two principal consequences for Montedison's strategy: on the one hand a stagnant world economy reduced the demand for chemicals and changed the structure of the supply, worsening Montedison's economic results and the debt/equity ratio. On the other hand inflation raised the cost of money, thus making indebtedness more expensive, and further worsening Montedison's results. Both these processes compelled Montedison to revise its plan.

The stagflation arising from the oil shock required, as in other countries, a quick and thorough response by chemical firms. However, in Italy, as we have observed, the response was delayed by oligopoly conflict, combined with political conflict among parties and party factions, and by an obsolete managerial culture which was too production-oriented and insufficiently market-oriented, too dependent on government financial support, too centralized and insufficiently internationalized. The worsening of company performance, together with changing political equilibria, brought about a change at the top of Montedison. Cefis left, and was replaced by Medici, a talented former diplomat who chose a low-profile strategy and steered the company out of political turmoil, thus laying the ground for the subsequent economic recovery. Medici was not a manager, but succeeded in creating some of the conditions necessary for effective management to take place.

Firstly, he contributed to an end to the oligopoly war in the chemical sector, with Montedison a winner, although an exhausted winner. In fact, Sir, Rumianca and Liquichimica were finally taken over by ENI, and Montedison was then able to start negotiations with ENI over a more rational division of product sectors and markets. The chemical war had consumed billions of lira of public funds, and paralysed corporate strategies at a time when foreign competitors were restructuring in order to cope with the new problems raised by international stagflation.

Secondly, with the help of Schimberni, who became vice-president for finance, Medici started the process of consolidating the company's financial autonomy by blocking attempts to take over Montedison, and by making Montedison's stock an appealing investment for private investors. Thirdly, he started to construct a managerial group consistent with the new strategy of reprivatization by appointing individuals according to managerial rather than political criteria. Fourthly, he revised the ambitious investment plan of the firm into a direction more in line with its financial capabilities, and started programmes of industrial restructuring (as in the area of fibres), making Montedison's strategy more selective and more business-oriented.

The three-year period of Medici's chairmanship was a transition period which, although it did not bring about substantial improvements in company results, laid the ground for the real recovery. In 1980, when Schimberni became president, he inherited a corporation still unbalanced and beset with problems, but with more peaceful relations with the major stakeholders (competitors, government agencies, unions, political parties) with a more unified management and more realistic goals. The most difficult tasks in terms of implementing adequate structures and improving organizational structures for the recovery were still ahead, and the financial situation was still gloomy, but the basic preconditions for recovery were present.

Montedison's recovery in the 1980s

Montedison's recovery was made possible by a set of factors, the most important of which were a more favourable situation in the international economy, a changing approach in government industrial policy and an effective leadership capable of implementing the new business strategy. Trends in the world economy created favourable conditions for Montedison's recovery in two major ways: on the one hand energy and raw material prices increased at a slower rate, and then diminished, thus contributing to a new profitability of chemical firms; on the other hand increased competition in the world market speeded up the process of internationalization of the Italian economy, which, in turn, required significant changes both in government economic policy and in the strategies of major corporations such as Montedison.

The increase of productive efficiency and of competitiveness became the dominant goals. With this objective in mind, as recommended in a white paper issued by the Ministry of State-owned Firms in 1980, the government made the cost of money more expensive; encouraged private shareholders to provide risk capital to substitute for state participation; did not oppose joint ventures and acquisitions of, and by, Italian firms; modified the inflationary impact of the wage indexation system and, in general, tried to increase the influence of market forces.

In the more favourable political and social atmosphere of the 1980s, Montedison achieved significant strategic changes under the leadership of Schimberni:

1 The rationalization of Montedison's presence in the chemical sector, and the reorganization of Montedison's non-chemical activities.
2 The reprivatization of ownership, and the search for a stable financial base.
3 Internationalization and the development of a market orientation.
4 The shift from a divisional structure to a holding structure.

We will not discuss all these strategies here; only those which are more relevant in terms of the relationship of Montedison with the government and other major stakeholders such as unions, state-controlled firms and the banking system. The first major strategy was *portfolio rationalization* – mostly, but not exclusively, in the chemical sector. We have seen that in the 1970s the increase in oil prices led to, among other things, an overcapacity crisis in basic chemicals, and a move towards fine chemicals for major world producers. Because of its troubled history Montedison was late among the major groups in reorienting its activities: at the beginning of the 1980s, 80 per cent of its production was still in basic chemicals. The first strategic goal was, thus, the striking of a new balance between basic and fine chemicals by selling or closing down non-competitive plants, restructuring what was retained in basic chemicals, and by concentrating new investments in fine chemicals. This was carried through with reference to specific businesses, for each of which a strategy to build, maintain or divest was worked out, based on the sector's attractiveness, and the unit's relative position on the market. The most difficult policy to be implemented was divestment, largely because of the unions' defence of existing jobs.

Montedison was helped by the Italian government's industrial policy. The 1981 Chemical Industry Plan called for a duopoly in the industry, with ENI as a commodity producer, and Montedison as the specialty company, and this implied a certain amount of swapping and closing plants so that each national champion could become competitive in Europe. In fact, a troubled ENI–Montedison capacity swap took place during the following three years, very similar in purpose to the ICI–BP agreement in 1982. The swap was complicated by the internal problems of both companies (ENI changed four chairmen in four years, and went through the negative experience of the joint venture with Occidental Petroleum; Montedison was still struggling to recover). A final agreement was, however, reached, which left ENI the Italian leader in low- and high-density polyethylene (LDPE and HDPE) and polyvinyl chloride (PVC), with 16 per cent, 8 per cent and 16 per cent of total European capacity, respectively; and Montedison leader in all forms of polypropylene (PP) and polystyrene (PS) with 12 per cent and 17 per cent of European capacity (1983 data).

Although the transfers of plants sometimes created problems (the most difficult one was Montedison's Sicilian complex at Priolo), and some critics argued that ENI as a state-controlled company was made to pay for the recovery of the privatized Montedison, the swaps, on the whole, helped to rationalize the Italian chemical industry, as the swap between ICI and BP Chemicals did in the UK. For Montedison it allowed the company to concentrate managerial energies and financial resources (including the money obtained from the sale of plants to ENI) on a more limited set of activities, thus strengthening Montedison's competitive position in the world market. This rationalization was, in a sense, the precondition for the strategy of internationalization, the major example of which has been the comprehensive deal with Hercules in pharmaceuticals and polypropylene. The evolution of the portfolio, as portrayed in Table 4.1 shows the progressive reduction of basic chemicals in favour of fine chemicals, health care and services.

Government's help was important, not only in helping the capacity swap with ENI, but in easing the disputes with the trade unions over the closing of plants and reduction of the labour force in order to increase profitability. Montedison could implement

Table 4.1 *Montedison's portfolio's evolution (percentage composition of sales, 50 per cent Himont excluded)*

	1982	1986	1991 (estimated)
Basic chemicals	41	31	23
Other chemicals	16	20	24
Health care	9	9	13
Energy	17	20	18
Services	16	17	22
Other activities	1	3	—

huge manpower cuts (about 20,000 workers in two years) because it was granted the special status which makes the state meet the costs of the state-subsidized insurance system (*cassa integrazione speciale*).

If one compares the relationship of Montedison with the political system in the 1970s and in the 1980s the difference is striking. Both the government and the unions seemed to acknowledge the needs and the 'reasons' of the firm, the government helping with industrial and special policies, but without interfering with managerial strategies; the unions showing flexibility and pragmatism in their defence of jobs.

Closely related to portfolio rationalization in the core chemicals business was the strategy of *reorganizing Montedison's non-chemical activities*. Montedison had, throughout its history, been involved in non-chemical activities, both in the industrial and in the service sector. In the early 1980s such activities accounted for about one-quarter of total output, but for a significantly smaller percentage of total investment. An objective of the new leadership was to define more clearly the role of these activities with reference to the core mission of the group. Some companies were either sold or prepared for being sold, such as Fidenza Vetraria and Vitrofil; others, such as the Standa chain, with 400 retail stores and 150 franchises, were restructured and made profitable; others were bought, such as the Nichols brokerage firm. In 1983 all service activities were grouped together in Iniziativa Montedison Terziario Avanzato (META), and listed on the Milan stock exchange. META includes a nationwide telecommunications and data-processing network, an integrated group of financial services and engineering and consulting firms, alongside Standa which became a major cash-provider. The idea of transforming a wide array of non-coordinated, non-chemical activities into an integrated group of activities in the service sector, traditionally rather inefficient in

Italy, was a major move in Montedison's new strategy. META has progressively increased its financial role in the group, and has been one of the most sought-after stocks on the Milan stock exchange.

Privatization of ownership

A further major strategy was the reprivatization of ownership. The poor performance of several state-controlled firms had created a political and intellectual climate favourable for Schimberni's bid for privatization. The operation was engineered by Mediobanca, Italy's most important investment bank. A consortium of prestigious private entrepreneurs (among them Agnelli, Pirelli and Orlando) acquired Gemina, a small finance company belonging to Montedison, as a vehicle for raising additional equity in the market. With these funds Gemina then bought the government's 17 per cent share of Montedison.

The privatization had consequences both for government policy and for Montedison's strategy. The former was significantly improved, shifting from direct involvement in Montedison, which in the 1970s had led to an improper politicization of business strategy, to indirect control through the setting of rules of the game and industrial policy planning in chemicals. The latter also improved, since locating new sources of finance became more difficult, and thus Montedison's management was compelled to work out an effective financial strategy.

Privatization had the effect of making banks more cautious about lending money in the absence of the implicit guarantee provided by the state entrepreneur. As we have seen, it was necessary for Montedison to rationalize its portfolio, and to improve its financial performance, both in order to reassure the banks and successfully to raise additional funds on the stock exchange. The effort succeeded, and led to Montedison's stocks significantly increasing their value on the Milan stock exchange, and to be quoted a few years later on major world stock exchanges.

This strategy was helped by the 'guarantee' provided by Mediobanca, and by the most prestigious Italian entrepreneurial families. However, at a later stage the control of this group proved too tight for a dynamic and aggressive management. In 1985–86 the idyllic cooperation among capitalists gave way to controversy, and a harsh battle for the control of Montedison was fought. These recent events can be interpreted – and have been so interpreted by Montedison's top management – as a case of conflict between financial ownership and managerial control, and as an effort to develop a public company free from the control of the traditional centres of economic power in Italy.

The dispute started when Montedison's top management decided to acquire control of Bi-Invest, a financial holding which was among the partners of Gemina, the finance company controlling Montedison, without letting the other partners of Gemina know about it. The result was that Gemina left the control syndicate of Montedison. Then Montedison's top management acquired control of a major insurance company, La Fondiaria, again through the purchase of shares on the market, and without the consent of Mediobanca and the traditional financial elite.

The attempt to create a public company in order to grant a large degree of freedom of decision-making to Montedison's management could hardly succeed in a financial system like that of contemporary Italy which, however rapidly developing, is still rather small. Thus, Montedison finally entered into the orbit of another major private Italian group, the Ferruzzi group, which considered Montedison both a good financial investment and an opportunity to develop its industrial business in the agrochemical sector.

Internationalization

Closely related to portfolio rationalization was the strategy of internationalization. A company the size of Montedison had to act as a global competitor, borrow from international banks and place its stocks in the major world stock exchanges, manufacture in foreign plants, buy and sell plants in the world market, and enter into joint ventures with other international firms. Montedison's management increasingly learnt to think and work out strategies as a global competitor with the help of managers who were recruited because of their international experience.

Montedison improved its performance in the international oligopoly game, striking a number of strategic alliances. In fibres the alliance with Monsanto (an American-based firm) made Montedison the largest acrylic producer in Europe, and one of the three leaders in European polyesters. In polypropylene, Montedison joined forces with Mitsui for process innovation. The major alliance was with Hercules. Hercules, as the world leader in polypropylene and the largest producer in the United States, was a major competitor of Montedison, but, at the same time, it was an ally in pharmaceuticals, through Adria, a US company created to sell Farmitalia (Carlo Erba's products, such as the famous anti-cancer drug, adriamycin) in the North American market. The result of the alliance was Himont, a joint venture between Montedison and Hercules, with both parents with equal ownership, and with US and European headquarters staffed, respectively, with the employees of Hercules and Montedison. Himont appears to be an example of effective interaction with competitors leading to mutual benefits.

Montedison's international involvement is also reflected in its need for a political capability outside Italy. Montedison's subsidiaries abroad have responsibilities for liaison with domestic government authorities. Moreover, the special importance attached by Montedison to its relations with the EEC is reflected by the important role it has played in CEFIC. Together with the development of a global strategy, an increasing emphasis was placed on marketing by the company, a changing orientation which had its effect both on Montedison's structure and its corporate culture.

The holding structure

The new strategies of portfolio rationalization and internationalization could not be based on the centralized and undifferentiated structure which had so far characterized Montedison. A basic organizational shift took place from a divisional structure to a holding structure which, through a few subholdings, controls a system of autonomous business units, each focused on appropriate product-market mixes. It is more an industrial than a financial holding company, since the functions performed at the centre are not only finance, but also research and development and strategic coordination. However, it allows decentralized responsibility for the management of autonomous sectors with different market requirements and different strategies. Fine chemicals, for instance, required a different strategy from that required by basic chemicals: more entrepreneurial, more sensitive to technological innovation, more able to deal with small-business and market niches.

Montedison's involvement in sectors such as pharmaceuticals, energy and services also created a need for a variety of strategies. Typical of Montedison's holding structure is the new portfolio formed by its non-chemical activities. On the basis of the principle that Montedison's mission was in chemicals, all other businesses were reframed with regard to that primary activity, but with a clear perception of their potentialities. Some firms were either sold or closed; those remaining were reorganized within the subholding, META. META provides services both for the other components of Montedison and for the market in a sector which is growing but is very competitive, and requires a quite different strategy from that employed in chemicals.

The actual structure of Montedison, as shown in Table 4.2, is divided into nine major divisions, plus two others where Montedison has up to 50 per cent of total shares. The creation of

Table 4.2 *Consolidated sales 1985 (in billions of lira), Montedison S.p.A.*

Petrochemicals and plastic materials 2601	Agrochemicals and fertilizers 945	Fibres 1083
Functional chemicals 660	Specialties 595	Health products 1293
Energy 3225	Consumer products and manufactured products 902	Services 2527
Shareholding up to 50%		
Polypropylene (resins and fibres) 1720	Paint and varnishes 236	

the new holding structure was paralleled by a shake-up of top management. Many of the newcomers were foreigners, and this contributed to the fostering of a more international and market-oriented company culture.[1]

ICI's development as a company

The fourth largest company by turnover in Britain, ICI has long enjoyed a status as the flagship of the British industrial economy. 'From the first, the company appeared to be regarded more as a public service than a powerful new private enterprise' (Kennedy, 1986: 50). Formed in 1926 as a counterpart to IG Farben, it was heavily involved in international cartelization in the prewar period which gave it a protected market in the British Empire and Commonwealth. The most important agreement was 'the Patents and Process Agreement, or "Grand Alliance", between Du Pont and ICI signed in 1929' (Mirow and Maurer, 1982: 131). In 1939 ICI was a party to no less than 'eight hundred agreements covering its chemical business' (Mirow and Maurer, 1982: 131). Throughout this period the company was under the autocratic leadership of Lord McGowan, who finally retired as chairman in 1950.

In the postwar period, ICI had to adjust to conditions in which markets were no longer protected by bilateral agreements with other firms, and in which there were a significant number of new, aggressive competitors in the form of the oil companies. 'Once the cartel had been broken up in 1947, ICI was at a severe disadvantage compared with Du Pont and even to the dismembered parts of IG Farben. . . . It took ICI twenty years after 1947 to develop significant European and American businesses and to reassert its position among the world leaders' (Stopford and Turner, 1985: 57). McGowan's successors as chairman tended to be distinguished scientists, but less experienced at developing new markets. However, in a period of growing demand for chemical products, ICI, with its technological strengths, was able to maximize its sales without developing much in the way of marketing skill.

Sir Paul Chambers, chairman from 1960 to 1968, did provide a 'break from the production and scientific emphasis of the past' (Channon, 1973: 143). In particular, 'Heavy capital investment was made in fibers, organics, plastics and fertilisers. . . . The company moved closer to the consumer-developing interests in pharmaceuticals, wallpaper and horticulture.' However, old habits die hard. It is clear from Pettigrew's study of the company that in many parts of the company the emphasis on technological excellence rather than commercial success, on allocating products rather than marketing them, was not easily displaced. Pettigrew quotes one manager in Mond Division who told him: 'Marketing was sales control literally, we told the customer how much he could have and how much he would have to pay for the privilege of having the product' (Pettigrew, 1985: 333).

In reacting to the strong leadership of Sir Paul Chambers, ICI moved towards a consensus style of management with the chairman (of whom there were four in the 1970s) limited by a series of checks and balances. ICI continued to make good profits, and reacted much too slowly to the changes in the world's economy which followed the first oil shock of 1973. In particular, it continued to invest heavily in large bulk chemicals on the assumption that growth would continue much as in the past. As Pettigrew notes (1985: 470), 'many of the core business issues of the 1970s, the heightening international competition, the sensing and responding to falling real growth rates for bulk plastics, what to do with maturing products, and how and when to develop new speciality products, were not faced up to'. As Pettigrew shows in his detailed study, the company had become characterized by conservatism, inertia, a difficulty in innovating and a paralysing system of consensual decision-making.

It was only after the second oil shock that the underlying weaknesses of ICI, particularly its overexposure in commodity chemicals and its dependence on home base sales, became apparent. ICI went into the red for the first time in its history in the second half of 1980. The election of (Sir) John Harvey-Jones as chairman in November 1981 reflected a recognition in ICI that the company needed a new approach to its problems. A somewhat unorthodox personality by the standards of ICI, but a man of undoubted ability who had worked up the company from humble beginnings as a work study officer without a degree, Harvey-Jones was a man of the times and for the times. His approach was not that of the 'first among equals' adopted by ICI chairmen of the late 1960s and 1970s (see Pettigrew, 1985: 383). He cut back ruthlessly where he thought it was necessary, regardless of company history; for example, he took ICI out of polyethylene, which it had invented. He built up ICI's strengths in specialty chemicals, using the company's financial strengths to make acquisitions where necessary (such as Beatrice Chemicals in the United States). These developments culminated in the division of ICI in 1987 'into two distinct arms; the traditional bulk commodity sectors and the speciality growth sectors . . . on which many of the company's hopes for future innovation must be pinned' (Kennedy, 1986: 178). As a chief executive in all but name, Harvey-Jones concentrated on two main responsibilities: planning and public affairs. These priorities reflected the need to turn the company round in the changed trading conditions of the 1980s and to handle issues arising in the external political environment.

Sir John Harvey-Jones's successor, Mr Denys Henderson, who took up office in April 1987, was a less flamboyant individual than his predecessor. Formerly head of ICI's paints division, Henderson was cast from a more orthodox ICI mould than Harvey-Jones. Whilst Harvey-Jones's style was 'the subject of some internal dissent from more conservative colleagues', Mr Henderson commented on appointment that his Scottish reticence 'might lead him to a less high profile' (*Financial Times*, 29 March 1986). It is also significant that, whereas ICI chairmen typically had about three years between appointment and retirement until the Harvey-Jones period, Denys Henderson had the prospect of an eight-year term in office. His selection can therefore be seen as a sign that ICI thinks that it is through its period of traumatic reorganization, and can settle down to a period of steady growth.

Internationalization

ICI has been a highly internationalized company ever since its formation. It was stated at the company's launch that 'in forming this combine we are doing something for the Empire' (Kennedy, 1986: 50). In the prewar period ICI's imperial market was protected by the cartelization agreements between the major international companies. By the 1960s,

> New thinking challenged the previous strategy of 'containing' their

> European and American competitors by means of large plants distributed across the Commonwealth and Latin America. World-class status and long-run competitiveness was seen as demanding that ICI learned how to compete with the leaders in their own backyards, just as Du Pont had challenged ICI in the 1950s. (Stopford and Turner, 1985: 75–6)

ICI expanded into Europe and the United States.

In recent years the company has undergone a transformation more fundamental than that brought about by investment overseas. The ownership and control structure of the company has changed, so that it is now in many respects an international rather than a British company. In 1983 ICI was listed on the New York Stock Exchange and 'By mid-1984, 16 per cent of its stock was owned by American shareholders – a significant dilution of its Britishness that could affect its future strategy' (Stopford and Turner, 1985: 130–1). By 1987 the board of ICI had Japanese, German and American directors. Their presence on the board was part of a consensus strategy of internationalization. In the company's 1984 annual report, (Sir) John Harvey-Jones noted:

> We are more geographically diverse than any other major chemical company and have steadily strengthened our position in Europe and North America, the two major chemical markets of the world. In the Far East . . . ICI already has a substantial presence and we are making a major effort to increase our activity in this region. The increasingly international nature of the ICI Group requires us to ensure that we have an international outlook and perspective.

One symbolic indication of this growing international outlook was that in 1987 the ICI board met outside Britain for the first time, in New York.

Restructuring through bilateral negotiation

The most important issue that ICI had to face in the 1980s was that of restructuring the company to take it away from areas where it was losing money, or making minimal profits, to activities where there were long-run possibilities for sustained profits. The reduction of loss-making activities has involved ICI in the negotiation of bilateral arrangements with major chemical companies throughout Western Europe.

Such a pattern of concertation between companies is not foreign to the chemical industry. The industry has considerable experience of joint ventures, bilateral agreements, etc. Indeed, Mirow and Maurer suggest that 'When a number of major chemical firms hope to control trade in a certain product, they generally do not join together in a formal cartel, but instead negotiate bilateral agreements with each other that achieve the same effect' (1982: 120). Any agreements concluded in Western Europe today do, of course, require the clearance of the European Commission, and although DG IV's general attitude towards restructuring efforts in the chemical industry may be described as a tolerant one (see Chapter 10, on overcapacity), it is certainly not implied that any of the agreements discussed here are similar to cartels in effect.

ICI's first major exchange was with BP. In essence, BP specialized in the production of polyethylene and ICI specialized in PVC, with each side transferring businesses and, although not as part of any agreement, making plant closures. This agreement was similar in purpose, although not in content, to that between ENI and Montedison. This emphasis on PVC made technological sense for ICI: it had access to substantial supplies of salt and chlorine, essential elements in the production of PVC, and a new PVC plant in West Germany.

However, PVC is one of the least commercially promising of the bulk polymers. It is particularly used in the construction industry, which has been depressed during the 1980s. There was substantial overcapacity; the number of PVC producers declined from twenty-four in 1980 to sixteen in 1985, but only 11 per cent of capacity was closed down. As one industry analyst commented, 'Demand is flat, prices are dangerously low, and Europe is awash with capacity.' ICI therefore sought a bilateral arrangement with another major PVC producer, Enichem of Italy. A new joint venture company was formed to take over the running of production, research and development and marketing of PVC for the two parent companies. The new company will be the market leader with a 25 per cent market share in Western Europe and may be able to firm up prices.

ICI has also engaged in a product swap with the French firm, Atochem. What all these arrangements show is an ability to engage in effective interaction with competitor firms to the mutual benefit of both companies. For its general strategy to succeed, ICI has to divest itself of loss-making businesses, and this cannot be done adequately through unilateral closures.

The position of the two companies in their national economies

In comparing the roles that ICI and Montedison perform in relation to external political actors such as national governments and the EEC, both directly and through the intermediation of associations, it is necessary to point to certain differences between the circumstances of the two companies. In particular, it is necessary to discuss their position in their respective chemical industries, and the differences in the two economies and political systems in which they operate.

ICI is the dominant company in the UK chemical industry. It is difficult to calculate its precise share of production in the industry, but one estimate is that it accounts for around 35 per cent of Britain's chemical output (Hardman, 1981: 11). Whatever the precise figure, there is no other chemical company of a comparable size. The next largest operation based in the UK is the Chemicals Division of BP, which had a turnover of £1900m in 1984, compared with £9909m for ICI. Montedison has a dual position with ENI in the Italian industry. Another contrast with Italy is that there is not a proliferation of smaller companies, and there are a significant number of medium-sized companies.

The UK chemical industry has been particularly affected by the economic conditions prevailing in the UK in the early 1980s. A massive appreciation of the exchange rate between 1978 and 1981, combined with the onset of world recession and a sharp rise in labour costs, severely undermined the ability of British manufacturing industry to compete in world markets. Italy is, of course, a late industrializer compared with Britain, and has not experienced the same kind of deindustrialization process. The different times and sequences of the industrialization process in the two countries have produced different domestic markets for the chemical industry, although neither domestic market is a particularly large one. However, the industrial structure in Italy is much more heterogeneous than in Britain. Alongside areas which are deindustrializing, such as Genoa, there are rapidly transforming industrial areas such as Turin, areas of fast tertiary growth such as Milan, and areas such as the Adriatic regions which are still undergoing a process of industrialization. This is a very different picture from the UK where 'some whole areas of industry seem . . . to have virtually gone from the United Kingdom, and it is unlikely, even though now they would compete, they would be opened up again' (Harvey-Jones, 1985: Q. 1343).

The question of dominance of one or two large corporations in a sector is even more evident in the Italian than the British case, because of the structure of the chemical sector. As the most up-to-date industrial census data (Istat 1983) show, local establishments with more than 200 employees account for 2.4 per cent of all establishments and for 55.2 per cent of manpower, while 66 per cent of establishments, mostly concentrated in basic chemicals, perfumes and cosmetics, have less than ten employees.

Studies by experts, government reports, and statements by

association representatives all stress the structural dualism of the Italian chemical industry. At the 1985 annual meeting of Federchimica, the president, Gianni Varasi, remarked that the Italian chemical industry was made up of a 'fragile system of firms', with two large groups, the one private and the other public; subsidiaries of some multinationals; a multitude of small firms, often very creative but condemned to a subservient role by the increasing globalization of the market; and a very thin segment of medium-sized firms, which in other countries plays a vital role in transferring know-how, and in stabilizing the system.

Political activities of the two companies

In the polarized context described in the preceding section, Montedison and ENI inevitably play a very important role in the definition of the issues to be discussed with major stakeholders, and in influencing the formation and implementation of government policy, although adopting a low-profile stance within the sectoral associations. An indirect but neat example of the key role played by Montedison and ENI in shaping associational policy and organizational structure was the formation in 1984 of Federchimica, a second-order association, made up of eleven subsectoral associations, and representing *both* private and state-controlled firms. The formation of Federchimica was favoured by a variety of factors, the most important of which were the organizational structure of Aschimici (its predecessor for private chemical firms) which allowed the transformation of product groups into first-level autonomous associations, and the similarity of policies and attitudes towards labour unions of Aschimici and ASAP (the association representing public chemical firms). Above all, the combination of private and state-controlled firms in the same association was made possible by the industrial agreement between the two large oligopolies, Montedison and ENI, which we described earlier, and which can be seen as a part of the whole process of rationalization of the chemical sector. The importance of the formation of Federchimica can be fully appreciated if one considers that it is the first and only instance of coexistence of private and state-controlled firms in the same business association since the 'historical' break of public enterprises away from Confindustria in 1957.

ICI was one of the first companies in Britain to develop a specialized government relations function. Its conduct of government relations is marked by considerable sophistication. It is also substantially involved in business associations, at the subsector level within the chemical industry, through the Chemical Industries

Association (and other national chemical associations in countries where it has plants, such as the VCI in Germany), within the CBI, and, of course, at European level through CEFIC (see Chapter 2 for further information about this organization). It could be regarded as a politically active company except in one important respect: it has a policy of not making donations to political parties.

There has, of course, been a historical tradition of companies making donations to political parties in Italy. In the 1980s the situation had changed; no large company would admit that they pay money to political parties. However, the overall environment in which chemical companies operate is still more politicized in Italy than in Britain. One factor is the presence of a state-controlled firm, ENI. Even in the case of Montedison, political connections are still stronger than in the British case. However, this is an area in which the Italian chemical industry is becoming less different from that of Britain.

How do large firms contribute to and benefit from business associations?

ICI is extensively involved in all the business associations in which it is entitled to be a member, ranging from subsector associations such as the Paintmakers' Association and the British Agrochemicals Association, through the Chemical Industries Association, to the CBI and CEFIC. It has provided two presidents of the CBI since its formation in 1965 and, like Montedison, it has provided a recent president of CEFIC.

However, although it has a *presence* on the key committees of all significant associations of which it is a member, it is careful not to dominate the associations. Indeed, it tends to be less well represented on key committees than its economic performance would justify. This point may be illustrated by a consideration of the nature of its involvement in the Chemical Industries Association. The president of the association is usually drawn from a firm other than ICI. Moreover, the numbers of employees it places on key committees are far fewer than it could claim given its importance in the industry. For example, in 1984, ICI had only one representative out of twenty-nine on the CIA's main decision-making body, the Council. Of the three other major committees it had two representatives (out of twenty-eight) on the Trade Affairs Board, two (out of twenty-four) on the Industrial Relations Board, and two (out of thirteen) on CISHEC, the environmental and health and safety committee. These numbers do not vary very much from year to year, except that there has been a tendency for ICI to reduce its representation over the years. Thus in 1975 it had three members (out of twenty-three) on the Council, and no less than seven (out of thirty-five) on the Trade Affairs Board.

This reduction in numbers should not be seen as indicating a loss of interest in business association activities, but rather a drawing back from a leadership position in the industry. A major company commented in interview: 'ICI takes less of high profile than it once did, urging other companies to play a more prominent role in speaking for the industry.' Another company commented: 'ICI don't want to be a leader of the industry, it has taken bullets in the past.' However, ICI's continuing commitment to business association activity is indicated by the quality of its representatives on key CIA committees in 1984:

Council – Executive Director
Industrial Relations Board – General Manager Personnel; Chief Negotiator
Trade Affairs Board – Company Distribution Adviser; Head of Public Affairs
CISHEC – Group Safety Services Manager; Group Environment Adviser

It could be counterproductive for a very large company to exert an influence on a business interest association proportionate to its economic weight. Such a stance could discourage smaller firms from joining the association, hence undermining its representative character, and counteracting the beneficial effects of the selective incentives cross-subsidized by the larger firms. Indeed, a dominant stance by a leading company could even deter second-rank companies from supporting the industry association. Hence, one finds that both ICI and Montedison adopt a 'low-profile' approach in relation to their respective business associations. As we have shown, this does not mean that they do not participate actively in the work of the associations. However, they try to ensure that they do not exert such an influence that the viewpoints of other firms are inadvertently downgraded in importance. Their role tends to be hegemonic rather than dominant. They try to build up a consensus on basic values, objectives and rules of the game, instead of exerting the power stemming from their greater size and resources.

But what do such large firms require from a business association? Certainly, they do not rely on its advisory services; often they make a greater contribution to the provision of such services than the benefits they extract from them. However, there are certain issues on which the industry needs to take a collective position; for example, trade policy questions or new environmental regulations. The expression of a viewpoint by an association can, in certain circumstances, give it a legitimacy in the eyes of government which it would otherwise lack. Association membership may also give a firm access to formal advisory committees and working parties within government, which it could not sit on as Firm X, but it could sit on as Firm X representing Association Y.

Direct relationships between large firms, government and unions

Governments often deal with associations, particularly when they wish to obtain a view from an industry as a whole, or to engage in formal consultations about a new policy proposal. However, they invariably develop direct relationships with large firms because of the importance of such firms in the economy. British governments have been particularly prone to consult directly with large firms. This tendency probably reflects the high degree of concentration in the British economy which is thought to be the most highly concentrated in the western world (Utton, 1982: 22). It is also a reaction to the weakness as intermediaries of many British business associations, combined with a lack of willingness on the part of government to intervene in the organization of what are seen as essentially voluntary organizations which must be left to order their own affairs.

Moreover, since the launching of the 'Industrial Strategy' in 1975, the emphasis in government's relations with industry has been on a 'bottom-up' approach which encourages the cultivation of direct contacts with companies. As a second permanent secretary in the Cabinet Office has explained:

> Traditionally . . . relationships between industry and government have been very much at the top, with a heavy reliance on industry's representative organisations such as the CBI, trade associations and so on. What I now detect is that, as compared with the 1960s, the era of the National Plan which was imposed from the top, and where sectoral implications were carefully worked out with the help of the trade associations, in the 1970s, there has been a greater realisation that government must also have contact with individual companies, to complement its relations with industry's collective representative bodies. That principle is now generally accepted and quite a lot of progress has been made. (Mueller, 1985: 105)

ICI's formation in 1926 occurred with government blessing. ICI accepted certain obligations, and government granted it a measure of favourable treatment in return:

> A distinctive feature of ICI was its willingness to accept quasi-public responsibilities, such as building the Billingham plant in the 1930s to produce fertilisers for an imperial market which never materialised.

> Conversely, the National Government was prepared to help ICI with its plans to produce fuel oil by hydrogenerating coal. (Turner, 1984: 12)

It should be emphasized that the company's relations with government have generally been good, regardless of the party in office. Indeed, at the time of the 1945–51 Labour government, one minister commented that it was easier to get cooperation out of ICI than out of the nationalized electricity industry (Williams, 1982: 133). The company has made extensive use of government subsidies, particularly regional aids for new investment, but has also benefited indirectly from other subsidies, such as those paid to farmers to use fertilizers.

However, its relations with the Thatcher government have been more strained than is usual. Its chairman is, of course, a Social Democrat in his personal life. When ICI reported a quarterly loss a reliable source recorded that ministers commented, 'Oh ICI – they're just as bureaucratic as the civil service' (Keegan, 1984: 148). Moreover, ICI failed in its efforts to persuade the government to amend legislation which it saw as having a discriminatory effect on its taxation treatment compared with other companies. In an unprecedented move ICI took a court action against the tax authorities, and ultimately won its case in the appeal court.

Nevertheless, it must be stressed that this episode did not disturb ICI's day-to-day relations with government. It continued to enjoy access to civil servants at all levels of the government machine, and to ministers up to prime ministerial level. Responsibility for public affairs work is taken up to the most senior levels within ICI, with Sir John Harvey-Jones treating it as one of his two personal responsibilities. Another member of the board was working for the government on two days a week in 1985 dealing with efficiency problems. There is also a senior manager who is designated as head of public affairs. He has a public affairs specialist working for him in Brussels, as well as a government relations manager in London. (The latter is usually a high-flying manager in mid-career, who is replaced every few years.)

However, it must be stressed that there is a company policy of devolution of responsibility for public affairs work to operating units of the company throughout the world. Hence, the chairman of ICI France is expected to monitor relevant political developments in France. Each division of ICI in the UK is expected to maintain contact with its sponsoring ministry; for example, the Agriculture Division with the Ministry of Agriculture, the pharmaceuticals division with the Department of Health and Social Security. ICI has plants in about 110 UK constituencies and it is the responsibility of local managers to keep in touch with 'plant MPs' and keep them informed on policies and practices. The government relations manager and his assistant advise operating divisions if they run into difficulties; for example, if they have to deal with ministries with which they are not familiar, or if legislation affecting their business is proceeding through Parliament. There is considerable respect among decision-makers for the way in which ICI organizes its government relations work. A civil servant commented in interview, 'ICI is a significant voice with a sophisticated government liaison unit and a view on a whole range of government matters, some wider than the chemical industry'.

The discussion in this section has necessarily concentrated on government, which is ICI's most important external non-commercial interlocutor. However, as ICI is a 'non-conforming' member of the CIA, it also conducts its own wage and conditions negotiations with the unions (including a staff association for foremen and supervisors). Montedison, by way of contrast, conducts its negotiations through an employers' association.

As we have argued in an earlier section, Montedison's relationship with government and the political system in general has passed through two distinct stages. The first stage was one of close links between the company's policies and state economic policy, both in the 'legitimate' sense that state social policy goals (such as full employment, industrialization of backward areas and development of strategic industries), were of foremost importance for orienting corporate strategy; and in the 'illegitimate' sense that managers made donations to political parties and party factions in exchange for government financial support and political appointments. In this situation the style of managers was political, not entrepreneurial, the criterion for evaluating performance was party allegiance rather than professional achievements, and corporate strategies were more important to political competition than to market competition. The second stage, started with Medici and continued by Schimberni, was a more autonomous course of action, in which government and major political parties remain important stakeholders of the company, but not a major source of strategic business decisions.

When do firms use associations and when do they deal direct with government?

It is difficult to generalize about the circumstances under which a large firm will advance its interests through direct representations on its own behalf to government and when it will work through a representative association. Indeed, many issues will be pursued

Table 4.3 *Methods of handling issues, in large firms*

Through firms	Through associations
Tax questions which discriminate between firms (including local tax issues)	General taxation policies which do not discriminate between firms
Negotiating government grants	Public relations on behalf of industry as a whole
Changes in the grant regime which adversely affect the company	Changes in the grant regime which affect the whole industry
Customs and trade policy implementation that directly affects the company	General trade policy questions: for example, GATT 'rounds', anti-dumping policy
Energy concessions for the company	General energy policy questions
Issues relating to individual plant closures	General discussion of overcapacity issues and rationalization
Direct relations with, or 'audits' by, environmental groups	Technical issues – health and safety, environmental, transport of hazardous substances

simultaneously through both channels. However, an attempt is made in Table 4.3 to distinguish between those issues which would be handled predominantly on an individual basis, and those which tend to be handled through collective intermediaries.

Conclusions

Although there are clear differences in the political environments faced by the two companies, particularly the greater politicization (especially in the past) of Italian industry, there are also important similarities. Both companies have to defend their interests in the political sphere – on an individual basis and in conjunction with other companies through subsector, sectoral, 'peak' and supranational associations. Although influential in such associations, both companies are careful not to dominate them, exerting a subtle but substantial influence on the conduct of their affairs.

The two companies are also becoming less different in various ways. Montedison no longer has a state shareholding and has freed itself of other political ties. Both companies have had deals with major oil companies, and have international strategies based on acquisition and alliance-building. The rules of the game of international competition thus become more important than differences stemming from the proximate environment. In general, one can say that constraints set by the rules of the game of international competition, and the increased emphasis in the 1980s on corporate efficiency, have become stronger than the differences stemming from different social, political and cultural environments.

These changes were not merely coincidental, but did reflect the operation of an underlying dynamic, although it should be noted that Montedison has moved closer to ICI in its methods of operation, rather than the reverse. In the case of ICI, increased competitive pressures in the international market could be dealt with by internal restructuring. In the case of Montedison a more fundamental reorientation of its outlook, and its relationship with the political system, were required. For Montedison the experience of internationalization occurred at a later stage than for ICI, and an appropriate response was postponed by the influence of non-economic variables. Over a relatively short period, from the early 1970s to the mid-1980s, the Italian business class acquired a new legitimacy, and business values became more central to Italian society. Thus, in the case of ICI, a reorientation of the corporate culture from within the enterprise allowed a new emphasis on marketing, whilst in the case of Montedison changes in the external environment were more important in facilitating the adoption of a new approach to the challenges presented by the international market.

Note

1. Since this chapter was written, two major developments have taken place in Montedison. First, the Ferruzzi group, which is one of the world's largest agribusiness multinationals, secured control of the company and started a process of restructuring in order to integrate Montedison into the Ferruzzi industrial–financial group. Second, the volume side of Montedison's chemical activities was transferred to a new company called Enimont, together with ENI's similar activities. This merger did not work well, partly because ENI is more sensitive to political pressure to maintain jobs. In September 1990, the Italian government allowed one partner to buy out the other, with Ferruzzi having first refusal on a price to be set by ENI. Finally ENI voted out Ferruzzi.

References

Channon, D.F. (1973) *The Structure and Strategy of British Enterprises*. London: Macmillan.

Hardman, R.J. (1981) 'The British chemical industry' (typescript). London: Jordan and Sons.

Harvey-Jones, Sir J. (1985) 'Memorandum submitted by Mr Harvey-Jones,' and answers to questions, in *House of Lords Report from the Select Committee on Overseas Trade*, Volume II: *Oral Evidence*. London: HMSO.

Keegan, W. (1984) *Mrs Thatcher's Economic Experiment*. Harmondsworth: Penguin.

Kennedy, C. (1986) *ICI: the Company That Changed Our Lives*. London: Hutchinson.

Martinelli, A. (1979) 'Organised business and Italian politics', *West European Politics*, 2: 67–88.

Martinelli, A., Chiesi, A. and Dalla Chiesa, N. (1981) *I grandi imprenditori italiani*. Milan: Feltrinelli, p. 242 and following.

Mirow, K. and Maurer, H. (1982) *Webs of Power: International Cartels and the World Economy*. Boston: Houghton Mifflin.

Morandi, L. and Pantini, G. (1982) *Dialogo sull'industria chimica*. Milan: Eras.

Mueller, A. (1985) 'A civil servant's view', in D. Englefield (ed.), *Today's Civil Service*. Harlow: Longman. pp. 100–8.

Pettigrew, A. (1985) *The Awakening Giant: Continuity and Change in ICI*. Oxford: Basil Blackwell.

Ranci, P. and Vaccà, S. (eds) (1979) *L'industria petrolchimica italiana*. Milan: Angelu.

Stopford, J. and Turner, L. (1985) *Britain and the Multinationals*. Chichester: Wiley.

Turner, J. (1984) 'The politics of business', in J. Turner (ed.), *Businessmen and Politics*. London: Heinemann. pp. 1–19.

Utton, M. (1982) *The Political Economy of Big Business*. Oxford: Martin Robertson.

Williams, P.M. (1982) *Hugh Gaitskell*. Oxford: Oxford University Press.

5

State Policy and the Chemical Industry in Western Europe

Pamela Adams

Chemicals has long been considered a key sector in government strategies for industrial development. In many countries, in fact, chemicals has been treated not simply as a *link* between different sectors, but as a locomotive of growth for the economy.

The economics of the chemical industry have changed dramatically over the course of the past two decades. During the 1950s and 1960s the industry witnessed a phase of growth and economic prosperity. Following the oil shocks of the early 1970s, chemical firms needed to scrap overcapacity, increase productivity and improve competitiveness. To do so firms sought to reduce costs, to diversify their activities and to develop international linkages in both research and production. National governments were confronted with political pressure to finance restructuring plans and, in many cases, to protect employment at the regional level. In more recent years, new problems concerning technological advancements and new areas of development have arisen. Growth in chemicals has become increasingly linked with innovations in other major industries including energy, health and agriculture. As a result the demands on the state have changed.

Government response to the problems in the chemical industry in the 1970s varied from direct intervention in the structure of the domestic industry, to the provision of indirect incentives for restructuring, to benign neglect. Judgements concerning the success of these strategies have been almost unanimous. As Bower summarizes:

> government interventions sometimes have led to a misguided development of industry and now may be slowing down or politicizing the response to changed economic circumstances. . . . Evidence shows that the governments of France, Italy and Japan bear significant responsibility for the poor economic condition that the companies experience today. . . . If reducing excess capacity and rationalizing of industry structure are taken as performance measures, private action by firms in the United States and the United Kingdom have produced the most impressive results. (Bower, 1985: 263–4)

The question now is how will governments respond to the new challenges posed by recent developments in the chemical industry and how successful will they be at encouraging industrial growth.

This chapter examines industrial policy towards the chemical industry in Britain, France, Germany and Italy with the limited goal of describing the objectives, contents and tools of government intervention over the past decades. It builds on the earlier examination of the economics of the chemical industry and provides a framework for further analyses of the extent to which state policies have protected existing levels of social and political consensus. Given the focus of the volume on the European industrial chemicals industry, only limited reference will be made to industrial policies in the United States or to policies in areas such as pharmaceuticals and biotechnology.

France

France has pursued the most direct and incisive policy in chemicals. This policy stems from a long tradition of state intervention in the economy that has provided the state administration with centralized tools to formulate long-term objectives and control over the instruments to implement them. In the postwar period this intervention has undergone three stages of development. First it assured that the market provided resources for industrial modernization, and actually managed this modernization in some sectors. Later the state sought to provide political stability and selective incentives for economic growth. Finally it has intervened to facilitate adjustment to a changed world economy. As Zysman notes, however, throughout all of these periods,

> whether the task has been to restructure industries in difficulty, to promote the French position in oil, or to support programs in electronics, alliances between bureaucrats, financiers and businessmen have played an important role. What made such alliances possible was not simply a system of cronyisms, but rather the nature of the marketplace ties between government, finance and corporations. (Zysman, 1983: 146)

French policy towards the chemical industry dates as far back as the 1920s and 1930s. To encourage self-sufficiency in chemicals policy-makers imposed tariff barriers of 25–30 per cent and blocked the import of various products. In contrast with countries such as Germany and Britain which favoured the formation of large, leading firms (respectively IG Farben and ICI), however, France did not seek to aggregate the interests of the many small companies that were operating in the sector into one giant champion.

This policy changed in the 1950s and 1960s following the formation of the European Economic Community as the government turned its attention to the concentration and rationalization of French industry. This strategy was rooted in a more general concern about the ability of France to compete in world markets given the low rates of investment and the relatively small size of French firms. Although chemicals was not identified as a key sector in the same terms as electronics, aircraft and nuclear energy, for which national champion companies were created, it was considered a vital sector for the economy (Vernon, 1974). Parts of the industry continued to be protected through tariff measures. The major thrust of state policy, however, involved regional policies to promote development and encouragement of fusions and joint ventures, especially between French firms.[1] This strategy culminated in 1969 when Progil and Pechiney-Saint Gobin were acquired by Rhone Poulenc, a private French firm with a long history in chemicals. This action, which was favoured by the government, resulted in the formation of a French company that was comparable in size to its major international competitors.

These efforts were extended in the early 1970s with the formulation of the 6th Five-Year Plan (1971–75). The first objective of the programme was to consolidate the structure of French industry and to create large, internationally competitive firms. In chemicals the government sought to rationalize production by eliminating the complex web of joint ventures that had been created as a result of previous policies, and by assigning specializations to domestic firms. The result was the formation of five major poles in the industry, three in the public sphere (CdF Chimie, EMC, SNEA) and two in the private sphere (Rhone Poulenc, PUK). The activities of these five firms accounted for 33 per cent of French chemical production by 1977 (versus only 16 per cent in 1951), indicating a greater level of concentration in the industry. In areas such as fertilizers the government decided that production should remain entirely in French hands, and imposed a division of labour between the major firms (APC, EMC and Rhone Poulenc). Similarly, CdF Chimie was encouraged to expand in paints and resins, while SANOFI (SNEA) was restructured and strengthened in pharmaceuticals and cosmetics. The second element of the 6th Five-Year Plan concerned an expansion of capacity and an increase in R&D activities in chemicals. State policy included an injection of approximately $4.7 billion into the chemical industry for investments.

The problems inherent in this plan, however, were exposed with the oil crisis and the economic recession of the mid-1970s. In France, as well as elsewhere in Europe, serious problems of overcapacity in chemicals arose. This was particularly true in petrochemicals, where surpluses resulted from both the rise in costs for raw materials and a reduction in demand (OECD, 1985). In addition, the mergers and rationalizations that had been imposed by the state administration were often ineffective in creating more efficient companies (Berger, 1981). Finally conflicts were inherent in policy that sought at the same time to build national productive competence through protectionist measures and to foster cooperation with foreign firms.

By the late 1970s the industry was subjected to a new set of government priorities. The official policy line focused attention on sectors such as telecommunications and aerospace, while developments in chemicals were seemingly left to market forces. In reality, however, the government directed considerable resources into keeping open unprofitable plants and maintaining production in areas that were already registering excess capacity. It was not until 1979, for example, that Rhone Poulenc received permission to begin its exit from synthetic fibres (Shaw and Shaw, 1983). This defensive line resulted from national security and foreign policy considerations, as well as from concerns about unemployment, especially in regions that were highly dependent on the chemical industry.

French policy towards chemicals was to be altered once more with the victory of Mitterand and the Socialist Party in the elections of 1981. The general goals of the Socialist platform were to reindustrialize France, to render French industry more competitive, to place stricter controls on the internationalization of the French economy, and to diversify the industrial base. Although the extent of policy change which followed this election should not be exaggerated, the Socialist Party did effect an extensive nationalization programme that changed the pattern of ownership in many of France's major industrial sectors. In chemicals the Socialist programme included the nationalization of Rhone Poulenc and the creation of a new division of labour between French chemical producers. Atochem (formed after Elf acquired parts of PCUK, ATO Chimie, and Chloe Chimie) was to specialize in base chemicals and plastics. CdF Chimie was concentrated in petrochemicals, and in particular fertilizers, while the new 'public' Rhone Poulenc was diversified in specialty areas such as pharmaceuticals, agrochemicals, fine chemicals and synthetic fibres. The objectives of this policy focused on competitiveness, but contained renewed emphasis on France's traditional desires for self-sufficiency.

Following this reorganization of the industry the public sector in France controlled 27 per cent of sales, 21 per cent of employment and 34 per cent of investments in chemicals. The state's presence in the sector, however, was much greater in basic chemicals than in secondary chemicals: respectively 48 per cent versus 16 per cent of sales, 50 per cent versus 18 per cent of employment, and 49 per cent versus 24 per cent of investments.[2]

Economic difficulties at both the domestic and international level, as well as political shifts within the government, however, forced a revision of French macroeconomic and industrial policy by 1983. As one observer noted, by 1985 the public enterprise system was drowning in red ink: the combined net losses, old and new, in the industrial sector increased by a total of 90.8 billion francs between 1981 and 1985. Restructuring was badly needed in many sectors, but implied a heavy burden on the state budget (Adams and Stoffaes, 1986). The radical change in policy that followed was in part reflected in the 9th Five-Year Plan (1984–88) in which the government shifted its attention away from concerns of employment and towards the modernization of industry and the promotion of science, technology and research. The plan was divided into twelve smaller programmes that concerned areas such as scientific education and vocational training (91.5 billion francs over the five-year period) and research and innovation (64.2 billion francs over the five years). Yet change was also evident in more subtle strategies that sought to encourage firms to exit from unprofitable lines of production and to expand cooperation with foreign multinationals.

In chemicals, the plan identified both fine chemicals (in particular Elf and Rhone Poulenc) and pharmaceuticals as two poles to support in order to increase the international competitiveness of French industry. This support consisted of both public financing and increases in regulated prices. The plan also identified a need to convert production away from basic chemicals. Finally a new and massive research initiative in biotechnology was launched, which involved firms in many areas of chemicals production.

In some cases the implementation of this new policy has proceeded smoothly. Rhone Poulenc, a company near bankruptcy at the time of nationalization, broke even in 1983 and registered a profit of $348 million in 1985. The company's acquisition of Union Carbide's agricultural division in 1986, moreover, symbolized its confidence to compete on an international scale. Similarly, Atochem came out of the nationalization process with a promising package of production units and was able to convert its losses of $188 million in 1983 into a profit of $3 million by 1985. Atochem has also been successful in breaking into international markets and increasing the presence of French chemical producers around the world. Although foreign direct investments in France were not mentioned in the government's original plan for restructuring the chemical industry, such investments became increasingly acceptable as cash for these projects was increasingly in short supply. Arco Chemicals, for example, was allowed to take control of Atochem's polyols business and the marketing of Atochem's production of propylene oxide and propylene glycol. Arco also built a plant for the production of propylene oxide and tertiary butyl alcohol at Fos-sur-Mer in southern France which was to serve as an outlet for part of Atochem's production.

The stated intention to rid policy of all remnants of nationalism, however, has also faced problems of implementation. CdF Chimie, for example, has registered substantial losses since the early 1980s. The troubled product sectors that the company inherited from the process of nationalization have been unable to generate the cash flow required for investments in higher value-added sectors. As one French report (Cours des Comptes) noted, moreover, 'CdF Chimie can't compensate for weaknesses upstream through privileged access to raw materials, nor downstream by significant presence on the specialty chemical market'. As a result the company's presidents have sought both government financing to cover losses and government permission to sell parts of the company or to form joint ventures with foreign enterprises in order to obtain financing for development. Although the new government of Jacques Chirac, elected in March 1986, had no *a priori* policy against foreign ownership in the chemical industry as long as production was maintained, and as long as it did not present a simple move to control market share, it would prefer not to see the company pass into foreign hands. At the same time the government has refused to give the large injections of financing that have been requested by the firm's presidents and has refused to force either Atochem or Rhone Poulenc to incorporate sectors of CdF Chimie's activities into their more profitable portfolios. Rather the government has followed an incremental policy of gearing up the company for sale by redistributing shareholdings and continuing to inject small amounts of funds to finance investments (*Chemical Week*, 17 July 1985).

In conclusion, French industrial policy towards the chemical industry has undergone constant changes throughout the postwar period. These changes reflect in part an attempt by the government to adapt policy to changes in the international economic environment. Such manoeuvres are difficult, however, in sectors such as chemicals, in which the variables of competition change so quickly and in which innovation is vital. Recent policy developments indicate a new opening to the role of international market forces in the industry. Yet signs also exist that traditional concerns about the power of the state and national sovereignty in key sectors persist. It is still unclear which of the two will eventually dominate.

Italy

In Italy as in France the state has intervened extensively in the development of the chemical industry. Both the objectives and the tools of this intervention, however, have differed in several important respects. These differences reflect contrasts in the political and administrative capacities of these two states to effect selective industrial policies. Italian policy-makers do not enjoy control over the flow of industrial finance to the same extent as their French counterparts, and rely mainly on direct funding (subsidies and low-interest loans). Few mechanisms exist, moreover, to weld the interests of the political parties and party factions that control the instruments and agencies of policy to facilitate the creation and successful implementation of state strategies for industrial development. Rather, the focus of negotiation is between single firms and the government. Such negotiation does not represent a link between business and the state, however, but reflects a relationship of 'diffident collusion' (Amato, 1972). The government seeks government support in an unprogrammed way. The only lines of communication between these actors consist of the periodic insertion of managers into the state structure to serve the interests of their firms.

During the 1950s and 1960s the policy objectives of France and Italy in chemicals were similar: to drive industrialization and to promote regional growth through the localization of large plants in underdeveloped areas. Italian policy-makers, however, did not develop the fear of foreign dominance that led the French to protect domestic producers and encourage national 'solutions'. Italy's policies met with success during the period of rapid economic expansion as chemicals became one of the country's major industries. They began to be questioned, however, in the wake of the oil crisis and economic recession in the mid-1970s. The Italian industry was the only chemical industry of a major industrial nation to register a net trade deficit during this period, due in part to the country's vulnerability to the oil price increases: over 85 per cent of Italy's energy requirements were supplied by foreign sources, and 87 per cent of these imports came from the Middle East and North Africa (Allen and Stevenson, 1974). Yet the industry's problems were rooted in past investment strategies which had been spurred by fragmented government authorizations and subsidized credit schemes. The result was a duplication of efforts (firms in the public and private sectors had invested in the same areas, thereby limiting their productive specialization) and an inefficient localization of plants (creating higher transportation costs and greater political obstacles to divestment decisions) (ENI, 1977; Banca d'Italia, 1977).

In 1977 a major new law (Law 675) for industrial restructuring and conversion was adopted by the Italian government and a new interministerial committee (CIPI) was established to oversee the implementation of industrial policy.[3] The general objectives of the new policy included an improvement in the trade balance through import substitution and export expansion; the conversion, modernization and technological development of industry in order to improve Italy's international competitiveness; and the rationalization of the use of energy and raw materials. The funds of Law 675 were to be distributed according to a number of sectoral plans that were to be drafted by the Ministry of Industry. In 1978 the first chemical programme was approved as part of this law. The sectoral programme that was to act as the cornerstone of government intervention in chemicals analysed three major areas of production: the ethylene cycle, artificial fibres, and fertilizers.[4] The broad objectives of the programme included the rationalization of production in base and petrochemicals, the development of activities in secondary chemicals and the maintenance of employment levels in threatened areas. Both political and administrative obstacles, however, impeded the implementation of this programme. Rather the direction of Italian policy was determined by a series of firm crises (Barattieri, 1984).

In 1978 a major Italian chemical company, Sir, was near bankruptcy. In an attempt to salvage a financial disaster the government created a consortium of banks to manage the reorganization and restructuring of the firm. At the same time another company, Liquichimica, suffered a crisis and was placed under the supervision of public authorities. The result was a change in the ownership patterns within the industry and an enlargement of the state sector to include parts of all major chemical firms.

Yet the government's stake in the future of the industry went further than its ownership of production. By the late 1970s the debts of chemical firms had assumed such dimensions that the failure of the industry threatened a crisis in the Italian financial system as well. In 1978, in fact, the net losses of three major firms (Anic, Sir, Montedison) amounted to almost 650 billion lira. These losses were translated into mounting debts and an inability to finance new investments. Such conditions were particularly serious given the urgent need for restructuring in Italian chemicals and the capital-intensive nature of the industry.

The ruling coalition during this politically delicate period, however, also faced strong pressures to protect employment.[5] As a result, restructuring was delayed as the public funds that were poured into the industry were also used to maintain operations in many unprofitable and obsolete plants, and even expand capacity in areas that were already suffering from overcapacity. In the case of artificial fibres the government threatened to cut off its subsidies to Montedison if Montefibre (its subsidiary) closed one of its plants in Piedmont, and if the company pulled out of its participation in the development of a major new complex in Sardinia (Shaw and Shaw, 1983: 151).

It was not until the early 1980s that the political climate in Italy began to change in favour of productive efficiency and international competitiveness. The signs of this change were evident in numerous areas of policy. First the government began to advocate the creation of two separate poles in chemicals: one centred within the public sector (ENI) and one within the private sector (Montedison). The long-term objectives of this plan were commercial, although the immediate task was to manage the crisis within the industry. The final agreement, that was worked out between the firms and CIPI, and approved by the government in 1982, created a division of labour between ENI and Montedison and represented a new basis for the state's intervention in chemicals.

With this agreement in hand the Italian government renewed its dedication to chemicals development. State assistance to the public sector increased from 432 billion lira in 1981 and 311 billion lira in 1982 to 2227 billion lira in 1983 (Artoni and Ravazzi, 1986: 39). Although not all of these funds were destined for base chemicals, the increase was intended in part to aid the rationalization of Enichem's cumbersome dowry. While Montedison was pegged to become the nation's major manufacturer of fine and specialty chemicals, in fact, Enichem was saddled with many unprofitable, debt-ridden operations in base chemicals. In recent years the public company has been allowed to reduce capacity and shut down plants in areas such as ethylene, low-density polyethylene and PVC. Government transfers to the public sector, however, are still being used to a large extent to subsidize industries at loss and buy out bankrupt private enterprises.

The private sector in chemicals, by contrast, has slowly moved away from government subsidies to private sources of financing. It should be noted that companies such as Montedison have received substantial aid in their restructuring efforts in the form of assistance for laid-off workers, which has also helped to appease the unions.

Chemical firms have also received funding for innovation from a new law adopted in 1982 directed towards the support of R&D outlays. As part of this policy the government established a number of national research programmes to be directed by the Ministry of Research and Technology: in chemicals these programmes are further subdivided into four major themes (pharmaceuticals, agricultural chemicals, fine chemicals and new cracking processes). It should be noted, however, that the sums available are small: the chemical industry was allocated approximately 184 billion lira (Momigliano, 1986; *Mondo Economico*, 1984, 1987). A final change in policy has been the increasing leniency of the state towards the internationalization of chemicals, including both joint ventures and foreign direct investments by Italian firms.

These changes have recently begun to reap positive results in the industry. According to the national federation of chemical companies, Federchimica, Italian chemical production grew by 1.4 per cent in 1986, capital spending increased by 12 per cent and R&D expenditures rose by 16.5 per cent. Commodity chemicals' share of total production, moreover, declined from 54 per cent in 1976 to 43.5 per cent in 1985, while specialty/fine chemicals' share passed from 46 per cent to 56.5 per cent. The two major producers have also begun to show signs of recovery, although their large debts continue to act as a drag on performance (see Figure 5.1). Enichem registered its first substantial profit in 1983 and has increased its R&D investments from $40 million in 1983 to $177 million in 1988. Similarly Montedison has moved out of the red and has initiated large programmes of investment and R&D (see Chapter 4, by Grant and Martinelli, in this volume). Both companies have also developed strategies to increase their presence in international markets, although they are closely controlled by politicians and union leaders who are concerned about the effects of these trends on employment and research and production levels in Italy. The changes in government policy towards chemicals, however, have also had disturbing effects on the industry. Foreign companies, which recognized the opportunities in the Italian market, rushed to fill the gaps left from the restructuring of the industry in the 1980s, mainly in specialties production. This inflow has made it difficult for the industry to reduce its trade deficit, which climbed to $4.6 billion in 1986.

More than in France, therefore, the chemical industry in Italy has represented a key to economic development which has included

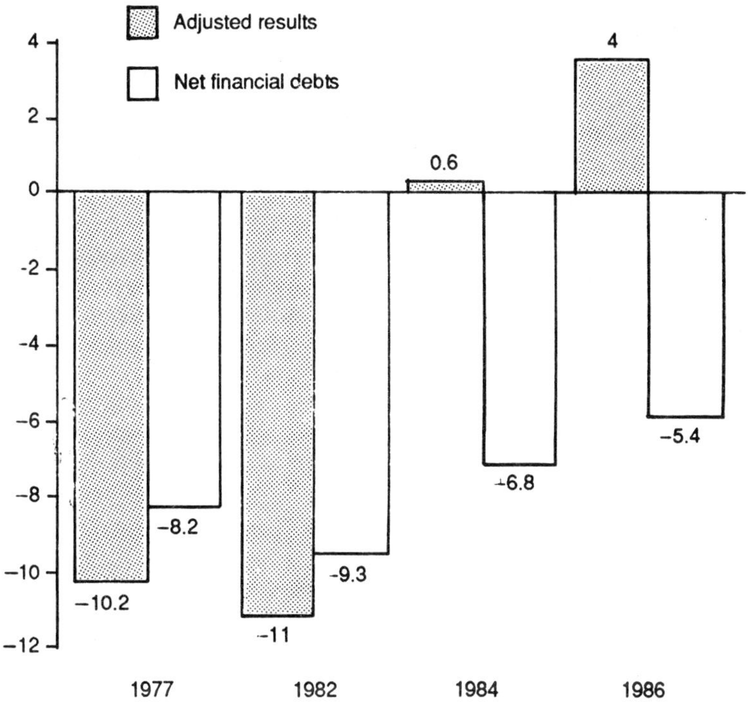

Figure 5.1 *Economic results for Italian chemical firms as a percentage of sales* (*Mediobanca*)

a large role for the state. The sheer size of the problems that arose in the 1970s forced the state to intervene to save the troubled industry. Unlike in France, this intervention did not involve an attempt to alter the structure of the industry or to create a national champion: still today the industry is characterized by a structural dualism that divides the few large producers from a host of smaller companies. Nor did the government seek to intervene directly in the general strategy of the firms. Rather, the state's role was limited to salvage measures to assure consensus. Although there are signs of a change in policy towards market forces, the size and concentration of the industry, and the extent of outstanding debts mean that the state will not be able to abandon old policies completely in the short run.

Britain

The British government has not been involved in either the construction or restructuring of the domestic chemical industry as directly as the French or Italian governments. This is due both to the character of the industrial policy network in Britain and the structure of the British chemical industry.

Britain has neither a long tradition of extensive government intervention in the economy nor an institutional framework to formulate and implement incisive, centralized industrial policies. Although the government has flirted with various forms of industrial policy, especially for depressed and developing regions, there

> has never been a uniform pattern of administration of economic and industrial policy throughout the UK. . . . The administration of industrial policy in Great Britain has been characterized by a high degree of institutional fragmentation both in the sense of there being a large number of institutions concerned with industrial policy at any one time, and in the sense of there being frequent changes over time in the departments and other public sector bodies concerned and in the degree of deconcentration and decentralization involved. (Hogwood, 1977: 4)

The major focus of the selective measures adopted in the postwar period, moreover, has been industries in crisis. This is not to say that advanced technology sectors have not been promoted by government programmes. Like France, in fact, Britain has maintained a commitment to a strong defence establishment; this is demonstrated by the size of the military's share of the total of national resources dedicated to industrial R&D. Sectors such as aircraft, nuclear power, and more recently electronics have also received government assistance for both defence and commercial purposes (Nelson, 1985). Yet like Italy, Britain did not develop a paranoia about dependence on foreign technology or foreign production, as long as it did not signify general economic inferiority. Political pressure to protect employment, especially in weak economic regions, moreover, has directed public resources away from government efforts to create an internationally competitive economy.

In the 1920s and 1930s cartel arrangements provided Britain's major chemical producer, Imperial Chemical Industries (ICI), with secure markets throughout the Commonwealth. The healthy profit margins maintained by the industry, as well as the close, informal relations established between ICI and the government in later years, helped to confine government intervention to indirect measures of assistance. Even during periods of economic difficulty chemical firms forged their own strategies in the absence of state intervention. In 1983, for example, a major 'swap' between ICI and British Petroleum (BP) occurred (Bower, 1985: 278–9). Both companies faced the need to restructure production and eliminate specific areas. The companies decided to swap rather than simply to eliminate their unprofitable areas, and prepared an agreement which was submitted to the Board of Trade and approved by the government. This agreement, like those that have occurred in West Germany, represented the independent efforts of firms to rationalize their activities.

Government policy-makers, however, have not totally ignored developments in chemicals. Although the extent of state intervention at the regional level was never as great as that in Italy or France during the 1960s and 1970s, chemical manufacturers did benefit from various regional development grants. It is estimated in fact that approximately 10 per cent of ICI's major plant investments were funded by the government.[6] In Scotland, moreover, oil and chemicals accounted for the largest share (21.6 per cent) of government development funds, and received more than £58 million in regional investment grants between 1975 and 1979 (Fraser and Orton, 1980).

The efficacy of state aid granted to the chemical industry, however, has remained an issue of debate. The focus of the debate has been the general objectives of government policy – to alleviate regional employment problems or to provide incentives for industrial restructuring. A study commissioned by Cleveland County, a major recipient of regional aid, found that the new investments financed by the development grants actually resulted in job losses. Employment declines, in fact, have been closely associated with technological innovation and restructuring in ICI's plants and with the rationalization of steel plants in the region. According to the authors, 'it is easy to argue that regional policy, as it affects Cleveland, is not in fact helping to alleviate regional problems but is an industrially oriented policy promoting restructuring in certain industries, with little or no employment pay-off' (Robinson and Storey, 1980: 17). Paradoxically, however, the report also indicated that chemical firms may have decided to invest abroad in the absence of this aid, thereby causing the complete closure of plants and an even greater level of job loss.

These and other forms of assistance to the industry were to be replaced in the 1970s as the British government attempted to imitate the successes of French planners and respond to the economic crisis facing the country. A new 'approach' to industrial policy was adopted in which information concerning the economy was to be fed upwards from industry and used in the formulation of macroeconomic plans. In chemicals, as well as other industries, sectoral working parties (SWP) were formed to design industrial strategies; the three SWP in chemicals concerned petrochemicals, specialty organics and pharmaceuticals. The efforts of these groups, however, met with little success. Although the plans facilitated informal communication between company and government representatives they did not directly result in formal changes in government policy towards the industry (Ilgen, 1983).

Rather, effective change in policy towards the chemical industry in the 1980s has come from the decline in regional aid (initiated by the Labour government in the late 1970s, but continued by the Conservatives after 1979) and the increase in support for R&D and in fiscal incentives for investment (Grant, 1982). The decrease in expenditures for regional policies was strongly opposed, in fact, by chemical producers. According to estimates of the British Chemical Association, for example, the industry stood to lose approximately $75 million of government financing by cuts in programmes designed to fund projects that would create new jobs in depressed regions.

By contrast the industry has had mixed feelings about the new R&D and investment incentives. Expenditure allocations to general industrial research and development by the Department of Trade and Industry (DTI) rose from £86 million in 1979–80 to £283 million in 1984–85, more than a threefold increase. The government also created new institutions to guide the government's efforts in R&D support, including the British Technology Group (formed in 1981, and responsible for transfer of technology through licensing and research grants) and the Science and Engineering Research Council (responsible for public research). Although 50 per cent of the funds allocated to the Support for Innovation Scheme and targeted for chemicals-related projects were earmarked for government-owned laboratories, private industry took little advantage of the remainder.

To spur industrial investment the Thatcher government also shifted attention away from discretionary assistance for selected firms to fiscal incentives. Such measures, however, have not always been applied in a neutral way. The Finance Act of 1982 represents a case in point. The Thatcher government was accused of allowing Shell and Esso to sell North Sea ethane to their chemical operators at below market prices. The government was found guilty on the basis that the tax allowances had the same effects as direct government subsidies, and therefore had needed approval of the European Community before being applied.

In conclusion, the British government has managed to avoid becoming directly involved in the development of the chemicals industry. The Thatcher administration has moved further towards the free-market model, reducing government assistance to the industry even further. Chemicals, however, remains a key sector in the British economy, which has been able to maintain profits and implement a major restructuring of the industry towards new products and new markets. Output in 1985, in fact, surpassed the 1979 peak by 8 per cent, and productivity has increased by 7.5 per cent per year. Chemicals also represents a major export earner for Britain. For these reasons the industry has the ear of the government, which may be used to assert political pressure for more favourable policies.

Germany

The chemical industry has played a key role in the economic growth of West Germany, providing the country with rich earnings from exports. In 1984 alone, exports increased by 16.5 per cent to more than $23 billion, accounting for more than half of total sales. The German industry is also one of the world's leaders in chemicals: three of the top four chemical firms in the world in terms of sales are German (Bayer, BASF, Hoechst).

The German federal government has not pursued a policy of direct intervention in the chemical industry except in times of war. The history of state assistance has in fact followed the pattern established at the very beginnings, when the state's role was limited to the provision of an institutional infrastructure for scientific training and industrial finance. Specific structural problems have traditionally been handled by informal market-sharing agreements between the firms themselves in the absence of state intervention.

IG Farben, in fact, was formed in 1925 by a merger between eight leading German companies, including Bayer and Hoechst. This new giant was to dominate the chemical industry. By 1929 it controlled over 100 plants and mines, employed almost 120,000, and grossed more than 1.4 billion reichsmarks (Hayes, 1987). At its height, IG Farben accounted for 35 per cent of German chemicals production, and was responsible for pushing Germany to the top of the list of exporting countries. This structure was dismantled after the war by the Allies and replaced by the creation of four separate companies. Paradoxically, perhaps, economic growth in Germany in recent decades has resulted in a stronger presence of German producers in international markets and an even greater level of concentration in the German chemical industry.

In the 1970s the competitive position of the German chemical industry was threatened by the oil price increases, the appreciation of the German mark and the rise in domestic labour costs. Again the firms took it upon themselves to design a strategy of informal rationalization. BASF retained its activities in bulk chemicals and fertilizers, while Bayer focused on specialty chemicals and Hoechst targeted synthetic fibres and pharmaceuticals. In turn, each firm agreed to reduce its interests in the strongholds of others.

The success of this informal agreement on the performance of the industry was soon evident. Despite recent currency fluctuations, falling oil prices and drops in sales volumes, all of the major German chemical producers have been able to maintain, and even increase, profits during the 1980s. The upswing in the market in 1984 allowed the Big Three to increase net profits substantially: Hoechst by 48.7 per cent to $434 million, Bayer by 55 per cent to $377 million, and BASF by 71 per cent to $288 million. These profits have in turn allowed the firms to increase investment and R&D outlays, which are so vital to their competitiveness, without recourse to public funds.

In the 1970s the government initiated a sharp rise in public support for R&D, which included the creation of a federal level ministry (the Ministry for Research and Technology, BMFT) to coordinate and expand government policies towards industrial R&D. The government's dedication to these policies is demonstrated by the rapid growth in the budget for this ministry: between 1972 and 1977 the BMFT's budget doubled, and by 1983 it had reached a total of 6.6 billion German marks. Still the share of company R&D expenditures financed by the government is relatively small in Germany: 15.7 per cent in 1984 versus 32.4 per cent in the United States, 22.4 per cent in France and 18.0 per cent in Italy (Malerba, 1988). In chemicals, government financing increased from 1 per cent of the total R&D outlays of firms in 1975 to only 4.1 per cent in 1983. The major part of public funding for industrial research is channelled instead into centres devoted to basic science research (i.e. Max Planck Institute) or applied research institutes financed in part by the government and in part by firms (i.e. the Association of Applied Research Institutes – the FHG).

The industry's prosperity, together with its international and 'independent' character, have also worked to limit chemical firms' reliance on the German financial system. As one recent analysis summarizes, foreign banks

> play an important role in the German chemicals industry, providing

> financial services for the external operations of German chemical firms. . . . German banks do play an important role in financing the chemical industry, but it does not appear to be as important as in some sectors. It is generally reckoned that the proportion of Eigenkapital (capital created by the firm) is twice as high in the chemical industry than the normal proportion in German industry. . . . The banks do, however, play a much more important role in relation to the smaller firms like Merck and Boehringer. Here they are not only an important source of finance, but also of investment advice. (Grant et al., 1987: 52–4)

Moreover, the leaders of the industry possess a strong feeling of group identification and a determination to run the industry on their own terms, limiting reliance on external capital.

Finally, chemical firms in Germany have not benefited from a strong regional policy. Some Länder governments have adopted regional aids to supplement the investment allowances granted by the federal administration. The sums in both instances, however, have been modest compared to those in other European countries. The restraint of the German government against intervening at the regional level to salvage jobs has been reinforced by the docile nature of the chemicals union – IG Chemie. In contrast with their counterparts in other countries, and even in other sectors within Germany, chemical workers have not resorted to strike tactics since 1971. Rather they have relied on negotiations to achieve gains in wages, job security and working conditions and have, in turn, accepted necessary cuts in employment.

Most economic indicators confirm that the success of the German chemicals industry should continue in the near future. German chemical producers spend generously on R&D: as a percentage of sales the R&D expenditures of the firms rank even with their major competitors in world markets (see Table 5.1). The large

Table 5.1 *R&D expenditures as a percentage of sales in major chemical firms*

	1980	1986
Bayer	4.3	5.1
Hoechst	4.3	5.6
BASF	3.5	3.4
Du Pont	3.5	4.2
Dow Chemical	3.1	5.4
Monsanto	3.1	7.6
ICI	3.7	3.8

Source: Company annual reports.

Table 5.2 *Total number of patent requests by firm, 1981–May 1987*

Bayer	4045
Hoechst	3128
BASF	3721
Du Pont	2289
Dow Chemical	2336
Monsanto	1385
ICI	1546

Source: Enichem.

number of patents requested by German firms (Table 5.2) indicates that these firms are also highly innovative. The strong presence of the industry in world markets, however, makes it especially vulnerable to currency fluctuations, changes in trade laws, and oscillations in prices for raw materials. In 1985, for example, exports were responsible for $27 billion of the industry's total sales of $60 billion. As one market analyst observed in 1987: 'the rising value of the mark – or the decline of the dollar – has put additional cost pressure on the West German companies, particularly the Big Three, which do about two-thirds of their manufacturing in Germany' (*Chemical Week*, 9 September 1987). German companies will therefore be paying relatively closer attention to the government's monetary and trade policy than to industrial policies in the near future.

Conclusions

In the 1980s the industrial policies of Western European governments towards the chemical industry have all moved closer to the free-market model of competition. In some countries this shift has entailed more radical change than in other countries. What is interesting to note, however, is that this shift in policy has not involved any major transformation of government institutions or in the structured arrangements of interest representation. Rather it reflects the changes that have occurred in the political coalitions that govern these countries. The French and Italian governments have slowly begun to reduce state subsidies to inefficient and unprofitable areas of chemicals production, which were kept open simply to maintain employment levels. Even in the absence of massive programmes of reprivatization, these governments have granted firms increasing autonomy to formulate independent investment strategies at both national and international levels. The road to a perfect free-market model, however, is long in these countries where government intervention in the industry has traditionally been more pervasive, and where the sunken costs of old policies prohibit an easy exit from protectionist measures. In Britain and Germany, by contrast, chemical firms implemented their own plans to rationalize production, while government intervention was restricted to indirect measures to spur investments and innovation. This was especially true following the rise to power of the Thatcher administration in Britain, when emphasis was placed on fiscal incentives and support for basic research in chemicals.

The political shifts that prompted the move away from dirigiste policies and salvage measures included external factors such as the successes of the Reagan administration in the United States, as well as internal factors such as the divisions within the national labour movements. Yet the specific characteristics of the chemical industry also played an important role in the process. The tasks and the limits of government policy, in fact, were defined in all cases by changes in the international economy of chemicals production and by the need for increasing amounts of capital for R&D. The oligopolistic structure of production, moreover, concentrated production in a few large firms in each of these European countries, and this concentration was reflected in the political representation of the industry.

Under such conditions it is interesting to analyse recent developments in government policy towards the chemical industry in areas such as biotechnology and new materials. Research and investments in these areas require large capital and personnel outlays by firms. State intervention therefore means that the government must find direct and/or indirect ways to provide a sufficient base and infrastructure for research, and to promote innovation in firms. Some experience in the technical aspects of innovation policy exists as a result of the development of policies towards industries such as microelectronics in the 1960s and 1970s. Yet the political aspects of biotechnology policy differ significantly from those of microelectronics. While the early policies towards microelectronics in Europe involved the promotion of relatively small firms in a new industry, in fact biotechnology represents the growth of a new area within an existing industry dominated by the production of a small number of large enterprises. Unlike in electronics, therefore, where political discretion was directed at a shift in public resources across industries within the broad context of national industrial policy, political choice in biotechnology involves a conflict between large and entrenched interests and small, new areas of expansion within the same industrial sector. This conflict presents chemicals, and its political agents, with an interesting dilemma to resolve. The problems that France and Italy faced in attempting to rationalize chemicals production demonstrate that the political strength of segments of an industry may be stronger than their market positions (Zysman, 1983). The problems that they now face in attempting to promote biotechnology illustrate the other side of the coin: the political strength of a segment of an industry may be weaker than its market position. In this sense the experience of the microelectronics industry is important (Adams and Orsenigo, 1988).

The conditions of the biotechnology industry vary widely throughout Europe. In Germany chemical firms began early to invest in new areas of research and production based on biotechnology. The firms themselves also promoted the development of government policy towards the industry which has been directed at financial support for basic research. In most other European countries firms have been slower to move in the direction of biotechnology, in part because they lack the necessary financial resources. The political lobby for government policy has not come from the industry, therefore, but from outside agents. In France the government, with its extensive capacity for autonomous action, was responsible for pushing the development of biotechnology. In Britain and Italy, by contrast, the political lobby for the industry had its roots in the scientific communities, alarmed by the lack of attention paid to biotechnology both by the government and by the firms. The development of government policies towards the biotechnology industry in all of these countries will now depend not only on the institutional arrangements of each national government, but on the political expression and political strength of the industry's needs. The stakes are especially high in this case given the renewed faith of many governments in the industry as a locomotive of economic growth.

Notes

1. The French government, especially under the Gaullists, often intervened to assure that important French firms in financial difficulty were not absorbed by foreign companies. The government sought at all times and at all costs to find a 'solution française' by encouraging French partnerships. The loss of Roussel-Uclaf to the German firm Hoechst, in fact, was regarded with great concern in government circles. See Morandi and Pantini, 1982.

2. These data exclude the energy sector. Basic chemicals includes synthetic fibres, while secondary chemicals includes pharmaceuticals. See Ministère du Redéployment Industriel et du Commerce Extérieur, 1984.

3. CIPI's members included the President of the Council of Ministers, and the Ministers of the Budget, the Treasury, Industry, Labour, State Holdings, and the Mezzogiorno. See Fornengo (1984).

4. The programme for fine chemicals was not approved until 1983.

5. The governing Christian Democratic party was forced, given their electoral strength, to bring the opposition (Italian Communist Party) into the government majority. The political convergence that followed, however, was centred in Parliament and focused on legislative procedures rather than on specific policy agreements. As a result, although important inroads were made in several policy areas, the coalition faced difficulties in solving the nation's major economic problems. The Christian Democrats succeeded eventually in coopting the left into a political programme to save employment with a minimum of loss of their own power. The Italian Communist Party (PCI) grossly underestimated 'how the routine processes of administration can be exploited by political forces bent on impeding or preventing change' (La Palombara, 1981). See also Colajanni, 1979. On the chemical industry see FULC, 1978; PCI, 1978.

6. Source: Interviews. The receipt of these funds was no doubt also the result of the close relations that had developed between ICI and the government administration in Britain. As one historian notes, there has existed 'a continual dialogue between I.C.I. and Government, sometimes with Government looking for action from I.C.I. . . . and sometimes I.C.I. seeking help from Government' (Reader, 1975; see also Chapter 4 in this volume, by Grant and Martinelli).

7. Between 1975/76 and 1977/78 Cleveland alone received more regional development grants (RDG) than Wales and almost 90 per cent of those of Scotland. RDG payments per capita were 14 times greater than the average level for assisted areas in Britain. Cleveland County estimates that between 80 and 90 per cent of investments subsidized by RDG between 1975 and 1978 went into Coal and Petroleum Products, Chemical and Allied, and Metal Manufacturing Industries. Such massive capital investments were also characteristic of the region in the late 1960s and early 1970s. See Robinson and Storey, 1980.

References

Adams, P. and Orsenigo, L. (1988) 'Tecnologie Emergenti e la politica industriale in Italia', *Rivista Trimestrale di Scienza dell'Amministrazione*, No. 2.

Adams, W.J. and Stoffaes, C. (eds) (1986) *French Industrial Policy*. Washington: Brookings Institution.

Allen, K.J. and Stevenson, A.A. (1974) *An Introduction to the Italian Economy*. London: Martin Robertson.

Amato, G. (1972) *Il governo dell'industria in Italia*. Bologna: Il Mulino.

Artoni, R. and Ravazzi, P. (1986) 'I trasferimenti statali all'industria', in R. Artoni and L. Pontarollo (eds), *Trasferimenti, domanda pubblica e sistema industriale*. Bologna: Il Mulino. pp. 15–45.

Banca d'Italia (1977) *Gli investimenti nell'industria chimica ed il loro finanziamento nel quinquennio 1971–1975*.

Barattieri, V. (1984) 'La ristrutturazione del settore chimico in Italia: un'analisi dei principali avvenimenti dal 1977–1983', *Rivista di Economia Politica*, 3 (3): 469–506.

Berger, S. (1981) 'Lame ducks and national champions', in W.C. Andrews and S. Hoffman (eds), *The Fifth Republic at Twenty*. Albany: State University at New York Press.

Bower, J. (1985) 'Restructuring petrochemicals: a comparative study of business and government strategy to deal with a declining sector of the economy', in G.C. Lodge and B.R. Scott (eds), *US Competitiveness in the World Economy*. Boston: Harvard Business School Press.

Calajanni, N. (1979) *Programmare in Italia*. Bari: De Donato.

ENI (1977) *Situazione e prospettive dell'industria chimica italiana*. Rome: ENI.

Fornengo, G. (1984) *Il problema della ristrutturazione industriale: la soluzione italiana*. Milan: Franco Angeli.

Fraser, N.A and Orton, I. (1980) 'Regional development grants: the Scottish experience 1975–1979', *Quarterly Report*, Fraser of Allender Institute, October.

FULC (1978) *Le proposte del sindicato per il piano chimico*. Rome: FULC [a trade union in the chemical sector].

Grant, W. (1982) *The Political Economy of Industrial Policy*. London: Butterworths.

Grant, W., Paterson, W. and Whitston, C. (1987) 'Government–industry relations in the chemical industry: an Anglo-German comparison', in S. Wilks and M. Wright (eds), *Comparative Government–Industry Relations*. Oxford: Clarendon Press. pp. 35–61.

Hayes, P. (1987) *Industry and Ideology: IG Farben in the Nazi Era*. Cambridge: Cambridge University Press.

Hogwood, B. (1977) 'Intergovernmental structures and industrial policy in the UK', paper presented at the ECPR Joint Session Workshops on Intergovernmental Policy-Making, Berlin.

Ilgen, T. (1983) 'Better living through chemistry: the chemical industry in the world economy', *International Organization*, 37 (4): 647–80.

La Palombara, J. (1981) 'Two steps forward, one step back: the PCI's struggle for legitimacy', in H. Penniman (ed.), *Italy at the Polls, 1979*. Washington, DC: American Enterprise Institute. p. 125.

Malerba, F. (1988) 'R&D growth in Italian industry in an international perspective', CESPRI Paper No. 5, Bocconi University, Milan, Italy.

Ministère du Redéployment Industriel et du Commerce Extérieur (1984) *Le secteur public industriel en 1983*. Paris.

Momigliano, F. (1986) *Le leggi della politica industriale in Italia*. Bologna: Il Mulino.

Mondo Economico, Monthly Reports on Chemical Industry, 31 May 1984 and 27 April 1987.

Morandi, L. and Pantini, G. (1982) *Dialogo sull'industria chimica*. Rome: Etas libri.

Nelson, R. (1985) *Policies in Support of High-technology Industries*. Washington, DC: American Enterprise Institute.

OECD (1985) *The Petrochemical Industry: Energy Aspects of Structural Change*. Paris: OECD.

PCI (1978) *Proposte per un programma dell'industria chimica*. Rome: Partito Comunista Italiano.

Reader, W. (1975) *Imperial Chemical Industries: a History*. London: Oxford University Press.

Robinson, J.F.F. and Storey, D.J. (1980) 'Employment change in manufacturing industry in Cleveland, 1965–1976', County of Cleveland Planning Department Report No. 176, May.

Shaw, R.W. and Shaw, S.A. (1983) 'Excess capacity and rationalisation in the West European synthetic fibres industry', *Journal of Industrial Economics*, 32 (2): 149–66.

Vernon, R. (1974) *Big Business and the State*. Cambridge, MA: Harvard University Press.

Wright, V. (1985) 'Socialism and the interdependent economy: industrial policy making under the Mitterand presidency', in V. Wright and H. Machin (eds), *Economic Policy and Policy-making under the Mitterand Presidency 1981–1984*. London: Frances Pinter.

Zysman, J. (1983) *Governments, Markets and Growth*. Ithaca, NY: Cornell University Press.

6

The Logic of Membership in Chemical Business Interests Associations

Antonio M. Chiesi

In the previous chapters we have called your attention to the relationships between business interest associations and governmental bodies, public opinion centres and other political actors. These relationships pertain to the so-called logic of influence (Schmitter and Streeck, 1981). Another type of social relationship is also important in ensuring organizational performance and levels of efficiency and efficacy of business interest associations in a given domain. As has been already noted (Martinelli et al., 1981), business associability has represented a serious theoretical problem since Olson's (1965) contribution. He points out that collective action in general represents a paradox for rational actors. In addition, unlike workers, businessmen exemplify individual behaviour within the context of market competition, where they hold the advantage given their power to allocate economic values. This structural position has been strengthened in ethical terms through the development of individually oriented competitive values that often regard collective action as collusive and unfair.

Nevertheless, business history shows that, in spite of the relevant role played by market forces, individualism and free competition, businessmen have tried to override their dependence on the 'invisible hand'. Their efforts have taken the forms of cartels, trusts, interlocking directorates and even the transformation of the market itself into what Chandler (1977) calls the 'visible hand'; that is, the internalization of market mechanisms into a rational organization through takeovers, mergers, combines, financial concentration, conglomerates and the like.

Seen in this light, the role that business associations play in the process of social regulation can be compared to the role played by other major regulatory mechanisms developed to cope with the increasing contradictions inherent in the market mechanisms (Lange and Regini, 1989). Both the state and organizational hierarchies replace the external impersonal mechanisms of the market with conscious regulation by political intervention. Although the 'state' can be analytically distinguished from 'business hierarchies' (the former is external to the business community and the latter internal and therefore controlled by entrepreneurs (Chiesi and Martinelli, 1988)), both have bureaucratic, authoritarian and centralized regulation. In addition, both enjoy an advantage over business associations, in that they can avoid spending their organizational resources to build internal consensus. This is because strategies are legitimized by virtue of the fact that they are made at the top. In contrast, in voluntary associations decisions have to be accepted by a majority of members and therefore are often the result of very complicated negotiations. An association has little real power to bring its dissident members in line. Associations try to over come this disadvantage, a result of their democratic structure, through largely formal participation mechanisms. They are helped in this effort by the fact that members traditionally show little interest in collective participation and tend to delegate their rights.

A number of studies devoted to the associational participation of entrepreneurs (Martinelli and Chiesi, 1988) have stressed the fact that, even among members, different levels of involvement can be found, ranging from complete disinterest to high degrees of commitment. The latter occur often without any reward or compensation to those who give up part of their workday towards such pursuits. Between these two extremes of disinterest and high commitment we find most members who simply stay informed about the association's activities through publications and leaflets, and occasionally by attending meetings. In short, the problem of participation in collective action by businessmen is two-fold. First of all it involves a firm's convenience in paying for a service whose possible benefits will be enjoyed by all firms, even those who have decided not to pay for membership. Secondly, from the viewpoint of the single businessman, there is the inconvenience of devoting time to collective action at the expense of his individual commitments.

Actually, as Olson (1965) points out, sufficiently large social groups, which pursue general benefits that everyone can enjoy (not divisible public goods) run the risk of failing their target, if it is supposed that single participants are rationally oriented by their own interest and therefore are able to understand that the most careful behaviour objectively suggests abstaining from collective action.[1]

Nevertheless a number of authors point out that the entrepreneur has been drawn to associative action by the need to counteract state intervention in the economy and union activities,[2] and has become a collective actor *malgré lui* (Schmitter and Streeck, 1981; Martinelli et al., 1981). Thus business associations must confront a problem of selective incentives to reward participation. These selective incentives are only a proportion of total associational output, and this proportion varies among different associations. Moreover, as I shall try to explain, members seem to appreciate different kinds of selective goods according to their economic characteristics.

Other scholars state that the premises, upon which Olson's economic theory is based, are exclusively utilitarian. They doubt that the simple model of an individual actor is realistic. When they try to explain collective action they relate it not only to the production of selective goods (an individual incentive to collective action that yields public goods) but also to the production of solidaristic goods. These show that collective action is dominated not only by the logic of utility, but also by the logic of collective identity-building (Pizzorno, 1983).

As the concept of collective identity allows Pizzorno to interpret the differences among political parties in different countries, in the field of business interest associations the concept of identity can be used to explain a number of associational differences and cleavages that cannot be accounted for by the specific economic interests of businessmen. In fact utility criteria are not able to explain, for instance, the development of confessional associations among Dutch entrepreneurs (van Woorden, 1984), the great variety of territorial groups without central coordination in the Canadian associational system (Coleman, 1985), the political cleavages among the artisan firms in Italy (Baglioni et al., 1983) or the ethnic and linguistic differences between the Spanish (Solé, 1985) and the Swiss entrepreneurs (Kriesi and Farago, 1985).

Trying to interpret the logic of membership in terms of collective identity is not, in my opinion, an illusory short cut in the explanation process, nor is it a rejection of the presumption that businessmen's actions are basically rational (Schmitter and Lanzalaco, 1988). In fact, as Pizzorno has shown (1983: 3), the concepts of utility imply a process in which each actor defines them within a community of reference, where they are acknowledged and reinforced. The community is able to foster collective action designed to pursue the social recognition of certain specifics. In other words, in addition to selective incentives the importance of solidarity and value incentives, as Etzioni calls them, should be taken into account to explain businessmen's participation to collective action.

In a recent survey, done on a sample of Italian entrepreneurs operating in five regions showing different models of industrialization (Martinelli and Chiesi, 1988), three main factors have been extracted that can summarize the range of motivations to associative action, according to the above theoretical hypothesis. The first factor relates to the representation and protection of the general interests of the firms in the political sphere and in confrontation with the unions. The second factor refers to the service activity supplied by the associations to their members. The importance of selective incentives is therefore emerging, especially that of consulting services in different fields, although they are represented by the second factor extraction.[3] The last factor clearly relates to a solidaristic dimension and stresses the importance of acquaintances and meetings among members that the association should develop, according to the respondents.

Furthermore, the survey shows that the idea of the association as supplier of collective goods and promoter of political activity is more often shared by members who more frequently attend the association's activities and are motivated by higher levels of participation and personal involvement.[4] The second factor shows a correlation with the size of the firm, both in terms of turnover and occupation, because small businessmen are more interested in the supply of services to meet their firms' needs. The third factor reveals a negative correlation with the number of years spent in the management of the firm. This result has been interpreted as follows: new businessmen, those who have less seniority, feel that the association, by developing socialization activities among members, contributes to the creation of a collective identity.

For these reasons I can state that solidaristic incentives play an important role in explaining business membership, but we can also stress that associations with specific religious or political ties (such as political party ties in Italy) enjoy a relative advantage in obtaining rewards for the category they represent. Moreover, it is plausible that associations of individuals are more sensitive to this dimension than associations of firms.

This chapter discusses and analyses the relationship between single firms and the collective organization of their interests, in the light of what has been defined 'logic of membership'. Specifically I will show the structure and problems of interest representation within the chemical sector, in a number of countries. Support for my arguments will be based on the results of the international research project coordinated by P. Schmitter and W. Streeck during the first half of the 1980s.

The process of business interest mediation and standardization is a strategic condition of any business interest association

(Martinelli, 1981). The management of this functional imperative is, however, only a necessary premise of collective action. In fact it depends on a kind of compromise between the logic of membership and the logic of influence. According to the former, the forms and means of representative organizations are dependent on the interests and the actual or potential behaviour of the firms in the domain. According to the latter, these characteristics largely stem from the associations' need to develop effective relationships with counterparts and institutional interlocutors.

I shall develop only the first of these two logics, analytically separating it from the second. I do so because my interest is not to analyse all associative behaviour, but only the relationship between members and interest associations. The logic of influence and the analysis of associative behaviour as a whole are discussed in other chapters of this book. My objective will be to discuss in depth the following points.

1 How do the specific characteristics of firms in the chemical sector (number, size, degree of productive integration, and so on) influence the interest organization?
2 How are the sectors' associations influenced by the fact that they belong to a larger associational system, which is sector unspecific and characterized by important differences at country level? I suggest that sector associability depends also on the logic of membership which is specific to each country.
3 The fact that the state, which is the main stakeholder in determining the logic of influence, also influences the logic of membership. This occurs, for example, whenever public firms decide to develop autonomous forms of collective representation, more or less separated from – or in competition with – those that represent private firms.
4 What are the factors that explain the domain definition of associations in the chemical sector? Empirical research has shown that the associability boundaries do not often coincide with the standard classification codes of production in this sector.
5 What is the role played by selective goods in promoting business collective action and what types of firms are more sensitive to them?
6 What are the associative contexts in which solidaristic goods acquire a certain importance?

Closing comments will be dedicated to the analysis of the relationship between degree of membership heterogeneity and organizational complexity between and within interest associations.

I shall also try to sum up the main problems that chemical associations have to face in managing diversity.

Sector associability

The chemical sector is divided into a number of different areas of specialization whose relative weight can affect the specific interests of the actors. This is why associations are usually articulated in several sections according to the kind of productions. In all the analysed countries[5] we can nevertheless identify a relatively stable associative structure, based on a large association which is often internally articulated in product-specific groups of members that take care of subsectoral problems. Sometimes this dominant association is a federation of minor subsectoral associations, such as in the UK (Grant, 1983) and in Italy after 1984 (Chiesi, 1985). Besides this main organization of interest representation we usually find associations for the producers of plastic materials that also include transformers and consumers. For this reason the interest representation in this subsector is often autonomous from the overall federation, as in the British, Swedish and Swiss cases.

There are also boundary areas that are related upstream to the productive processes of the chemical sector, as in the case of energy, or downstream, as in the case of man-made fibres and building materials. These can join either the chemical associational system or one from a bordering sector. This alternative can be in part explained by the specific development of the chemical sector in each country. For instance, in Italy the central role played by the petrochemical industry explains the inclusion of oil refining and transformation of hydrocarbons into the representation system of the chemical sector. This creates a kind of overlap between chemical and energy associations. Also, in some situations, the associations representing the interests of man-made fibre producers are included in the chemical sector; in others they belong to the textile associational system. Affiliation depends on the relative integration of this type of production downstream through chemical activities or upstream through the textile industry.

Another example is the pharmaceutical sector, which in most cases develops a separate association system, due to its specific problems caused by state control and authorization of specialties, price regulation, and to the sector class relations with public health systems. In this case the associational structure can be easily explained by turning to the logic of influence. Even so, in places such as Switzerland, where the major firms are integrated both in the chemical and pharmaceutical sectors, the relative associational system comprises both sectors. In other cases, such as Great Britain, the specificity of the pharmaceutical firms is acknowledged at the level of interest organization with a separate association. This, however, is part of a federation covering the chemical sector at large (Sargent, 1983).

Compared with other industries, the diversity of interests in the chemical sector is attenuated by the fact that there are relatively few actors in this domain, due to the considerable industrial concentration. The subsector of industrial chemicals often plays a major role in the formation of total turnover. Industrial chemical products are not differentiated among different suppliers, because of technology standardization. The production of basic chemicals is dominated by plant concentration, because of the increasing economies of scale and the rapidly growing size of the optimal plants. The technology of ethylene production is pivotal, requiring vertical integration of different processes around the ethylene cycle (Ranci and Vaccà, 1979).

Also in fine chemicals there is a strong tendency to oligopoly, for reasons ranging from high entry barriers to the need to diversify into a wide range of products and the concomitant importance of marketing. There is need for large international distribution networks, to expedite the sale of huge lots of diversified product lines. In this environment a manager's main problem shifts from production to research and sales.

For both basic and fine chemicals the result is an oligopolistic structure which is all the more important, the more developed the sector in a given country. For these reasons we find a limited number of firms in the sector[6] and a considerable share of the total output supplied by a few oligopolies often operating on multinational scale.[7]

The associative structure of the sector is influenced first of all by the kind of associable firms, as evaluated by total turnover and employment. In fact the logic of membership is different according to the firm's size, both in absolute terms and in relation to the sector. The articulation of the sector, the degree of integration among plants and the amount of exchange between the producers are also important.

As far as the distribution of firm size is concerned, it is important to stress the fact that the oligopolies can pursue different strategies from those of small firms. Under certain conditions the former's behaviour appears at first counterintuitive to Olson's thesis. The analysis of the representation of employers' interests in the Italian chemical sector (Chiesi, 1985) shows how often large firms take initiatives which usurp the official functions of the collective organization of the sector. This behaviour is obvious in the pursuit of individual advantages, more easily obtainable by large firms which can present themselves with authority before the unions, and which possess sufficient large-scale economies to support a lobbying structure or an autonomous technical–political apparatus during bargaining rounds. On the contrary, small firms must lean on the associations for such services, which are provided as selective incentives.

In contrast, the production of public goods is quite another situation, as in the case of drawing up collective agreements. Here, according to Olson's well-known rule, in a scenario with many players, the actor of small size can find a relative advantage in reaching agreements outside, or even against, the platform of the association.[8] Larger actors are constrained to follow association policy in a more orthodox way, because of their greater political visibility. On the other hand, the large actor is more able to influence the positions of the association. In sum, the logic of collective action of business interests appears more influenced by firm size than by the number of actors. Therefore the associations find it rather difficult to mediate between the interests of small and large firms, especially whenever the associative domain is dominated by a sharp dualism in firm size.

Turning to the articulation and degree of productive integration of the main producers in the sector, two main strategies are identified here, given that the chemical sector is dominated by a technology conducive to high upstream and downstream integration. Either the oligopolies can adopt a strategy oriented to cover most of the process, as in Italy (especially during the 1970s) or to confine themselves to initial, intermediate or final segments of the process, sharing the different tasks and coming to an agreement on mutual integration and coordination mechanisms, as in the case of Germany. The relative position of the chemical firms in the reciprocal division of labour has heavily affected the sector crisis during the 1970s, determining the different degrees of exposure to overcapacity and price increases and different response capacity.

English and German oligopolies, for instance, held a significant position in fine chemicals, in terms of proportion of total turnover; therefore they were less exposed to the oil crisis of the 1970s. In the meantime upstream productions, ranging from oil refining to the production of basic plastic materials, were controlled by the oil companies, and the losses resulting from overproduction in petrochemicals were counterbalanced by the revenues generated by oil extraction. On the contrary, in Italy the main chemical firms have traditionally been not only producers of basic chemicals, but also involved in oil refining and fuel distribution, although they did not have the advantage of revenues coming from direct control of the oil fields. This situation, which explains the magnitude of the sector's crisis in Italy, also helps to explain a certain degree of associative overlap between chemical and energy sectors.

Chemical associations as part of a larger system

The specific details of sector bargaining are less evident in a context of centralized industrial relations. In this case employers' associations in the chemical sector tend to take on the organizational features of their country's system of interest representation. Because the chemical associations are in turn usually affiliated to a peak association, representing trade interests, the larger system can also affect the chemical associations at this level, although to a lesser extent. Those characteristics of sectoral associations that cannot be explained by the firms represented can be understood by analysing the associational system at large. The former, that is the sectoral associations, are drawn to conform their organizational structure to that of the latter.

First there is a correlation between interest specialization at sector level and at country level. If the functional specialization between trade and employers' associations is adopted at the level of peak associations, we find this specialization also at sector level. Moreover the degree of comprehensiveness of interest representation at the level of the larger system is often reproduced at the level of the chemical sector. For example, factors external to the sector, such as ownership differences or political party allegiances, may result in sectoral cleavages. The more the sector associational system is integrated into the larger one at country level, the stronger the impact of sector-unspecific factors on interest representation in the chemical sector. I present two opposing examples, one British the other Swedish.

In the UK, chemical associations are embedded in a system of interest representation which is relatively inclusive because it is part of the Confederation of British Industries,[9] which encompasses private as well as public firms; small and large businesses; and industrial, trade and service sectors. The CBI was created in 1965 as an organizational merger between the once separate trade and employers' functions. In spite of the dominant position of industrial firms (in terms of fees and number of affiliates) the CBI is the peak of a rather loose associative system, which leaves great autonomy to member associations.

These characteristics are evident also in the chemical sector, where the Chemical Industries Association follows the general example unifying trade and employers' functions. As far as the latter are concerned, some members do not conform to the collective bargaining strategies of the association, as a result of problems originating in the period between the two world wars. ICI decided at that time not to conform to the policy of the association. This demonstrates that the lack of separation between the two functions can shift difficulties in interest mediation from employers' to trade interests. Moreover this is an example of how differences in members' size create problems in representing different interests, as discussed earlier. In this case history demonstrates how a single big firm with a dominant position can reduce the legitimacy of its corresponding association.

Sometimes the industrial relations system, of which business interest associations form a part, centralizes collective bargaining at national level. Consequently, we find less differentiated organizations of employers' interest at single-sector level. Because the representation of trade interests implies the recognition of many specifics, through the specialization of the associative structure, the potential contradiction between the two needs is worked out by organizing two functionally separate systems of interest representation. This is the case in Sweden where the overall associative system is less comprehensive than the British. SAF organizes only employers' interests in industry, commerce and services, while public firms, accounting for more than 30 per cent of total employment, have a separate association. The same holds true for the cooperative system, banks and the publishing sector. In sum, the Swedish system is characterized by strong political and ideological cleavages in the associations, operating in a context of strong trade unionism[10] and strong centralization of labour relations – also because of the role played by the indemnity funds (Skogh, 1984).

The chemical sector is influenced by the general conditions of the associative system of which it is a part, because we find in it a reproduction of the clear division between trade and employers' associations. The latter are sector-unspecific in Sweden and therefore in the chemical sector we find the same vertical tripartition (according to the kind of ownership) seen in the overall system. This lack of system integration at the level of labour relations does not prevent a higher degree of unity in the field of trade interests. KiK is the federation of associations within the chemical sector, and is in turn a member of IF, the sector-unspecific trade association at country level. Therefore at the level of employers' interests associative heterogeneity is due to political and ideological factors originating outside the sector. In the representation of trade interests the heterogeneity is endogenous and stresses subsectoral differences between domestic producers on the one side and importers and foreign multinationals on the other. Furthermore, as in Italy we find a considerable degree of cooperation between sector federation and subsectoral associations that do not join the associational system at the third level, that is the level of sector-unspecific peak association.

In the German case we also find a clear partition between trade and employers' associations. At sector level each associative system is therefore affiliated in turn to a peak sector-unspecific association at federal level. The associations of the chemical sector reflect this clear-cut functional partition. BAVC is a federation of associations and is in turn affiliated with BDA. It deals with labour relations problems and links thirteen territorial associations. Trade interests are represented by another federation that affiliates twenty-three subsectoral associations.

The associative system of the Dutch chemical sector is less systematic. It is part of a larger system of peak associations where the confessional cleavage, together with the territorial one, is emerging as a main feature. VNO is the most important federation and joins together the sectoral organizations. In addition there is NCW, the Union of Christian Entrepreneurs, which is the result of a merger between the previous confessional associations of Catholic and Protestant origin. It has a network of territorial associations that follow the diocesan areas of influence. These two organizations have formed a consulting and coordination body, the RNWV. The system is therefore characterized by a sectoral dimension of representation through VNO and by a territorial dimension represented by NCW. In contrast the functional partition between trade and employers' interest representation does not operate at the inter-associational level, since the two functions were taken over by VNO in 1968, after the merger of two previous associations.

The interest representation of the Dutch chemical sector is organized by a mixed association (VNCI),[11] into the larger system coordinated by VNO. However, VNCI does not represent the whole sector, because some firms producing synthetic fertilizers, resins and such items are represented by associations affiliated to AWV, which is an inter-sectoral second level federation directly affiliated to VNO. State ownership is heavily present in the chemical sector through DSM, accounting for a quarter of total employment. Unlike other countries, as discussed in the following, the heavy presence of the state in the sector does not seem to affect the organization of interest representation.

The Italian case, compared with the British and Swedish ones, is characterized by a more important role played by the territorial organization of interest representation (Lanzalaco, 1988). Nevertheless this element is certainly less important in the sector under study, and has minor consequences on its associative articulation, because the small firms are concentrated in the northern regions, especially in Lombardy, and because the large petrochemical plants settled in southern regions depend on their headquarters in the north. The territorial articulation of the associations keeps a certain importance where the problems of small firms have to be dealt with, and whenever industrial relations become markedly decentralized, as in the period between the 1970s and 1980s. This was due to the restructuring process and the management of redundancies in the large plants, producing basic chemicals in southern regions. The two main cleavages in business interest representation relate to the partition between large and small firms and between state and private ownership. Although the latter partition is present also in other countries, it is crucial in the Italian chemical sector.

The logic of representation according to the kind of product is not rigidly defined and exclusive. Especially the larger associations tend to include in their domain economic activities in sectors that relate to chemicals. This implies a potential area of representation larger than the chemical sector as defined by statistical criteria. Another consequence is that competition among associations can arise for the control of membership pertaining to overlapping domains. This lack of a systematic definition of boundaries between sectoral associations has been made less confusing since the mid-1980s, when the larger association (Aschimici) was transformed into a federation (Federchimica) with a leadership in sectoral interest representation.[12] The lack of similarity between associative domains and sector boundaries underscores the fact that sectoral criteria, compared with ownership and firm size, have little importance in determining interest representation.

If we focus our attention on small businesses, the associational system appears strongly influenced by the political sphere and follows the divisions of the main parties. I have already discussed the fact that the functional division of labour between trade and employers' associations is hardly present. In the Italian case the same association represents both interests and develops an internal organization which carries out the two functions. The chemical sector also follows the rule because Federchimica has always conformed to the organizational suggestions coming from Confindustria. Nevertheless an important exception should be noted. It can be explained with reference to the goals that the association of state-owned firms in the chemical sector have historically followed.

ASAP is the compulsory association of firms held by the state agency ENI, operating mainly in the energy, chemical and other manufacturing sectors. Especially during the 1960s ASAP developed only as an employers' association to break the monopoly of business interest representation that Confindustria had held in the field of industrial relations. This policy was devised by the Christian Democratic party to open a new strategy towards the unions (Martinelli, 1981). In the following section these events will be analysed more deeply, to demonstrate how state intervention can affect the logic of membership. Here it is sufficient to stress the fact that, in recent years, state-owned and private capital have begun to coordinate policies, notably in the management of redundancies and the restructuring of basic chemical productions. This has produced a de facto functional separation, an unusual case in the Italian representation system. In fact, while ASAP is specialized as an employers' association, the firms of ENI are now members of Federchimica in terms of their economic interests.[13]

The total number of associations in the sector is higher than the average for the other countries studied. This is mainly explained by the presence of a large number of artisanal territorial associations. They do not exert a significant role in the sector as such; so it is important to stress the uneven role played by the different associations. Federchimica, ASAP and Unione Petrolifera play a prominent role, compared to the above-mentioned associations. Their importance is also vastly superior to the territorial sections of the confederation of small firms (Confapi), where membership is drawn from similarly sized firms, whose existence is recognized and protected by the political system.

Furthermore we can cite the Austrian case, demonstrating how the articulation and comprehensiveness of the associative system depends on the status enjoyed by voluntary associations as part of a compulsory system of representation. The explanation of the case lies in the fact that chambers of commerce play a central role as official partners of the relevant state agencies and as exclusive negotiation bodies with the unions. Therefore a voluntary system of business associations can only have a residual function, both as trade and as employers' associations, and it is limited to dealing with issues not officially allowed by law or ones with which the chambers are not suited to deal (Traxler, 1983). Austrian voluntary associations are therefore task-specific and do not compete among themselves.

This marginal position explains: the lack of hierarchical articulation among the five voluntary associations that operate in the chemical sector (there is not a sector federation coordinating subsectoral associations, as in most other countries); the evident overlap in the domain of the three subsectoral associations (without originating competition among them); and the lack of coincidence between associative domain and sector boundaries (that is, the associations which are relevant from our point of view also join man-made fibres and textile industries in general, thereby reducing the sectoral specificity of the domain.)

The influence of state intervention on the logic of membership

The state is the main interlocutor of the chemical business associations in the area of economic interests, and also plays a relevant role in industrial relations. Therefore it is an important actor capable of determining the modes of the logic of influence. Nevertheless, this fact should not reduce the important role played by the state both directly and indirectly in the sphere of the logic of membership. Indirectly, legislation recognizes specific legal advantages to different kinds of firms, differentiating actors according to tax facilities, credit opportunities, vocational training aids and labour costs relief. This, for instance, is the case of the artisanal sector in two countries in which the respective associative systems at large show relevant differences: the German mixed system of chambers of artisans and guilds (Streeck, 1986) and the Italian system of party-oriented artisanal associations (Baglioni et al., 1983). In this field of interest representation the associative action is undoubtedly important in requesting specific benefits to the state. But it is also true that the state intervention acts directly on the membership cleavages that in turn can influence the associability criteria.

Direct state intervention in the economy is often accomplished through public ownership of firms. These can establish relationships of competition or cooperation with private firms and originate different patterns of interest representation, showing different degrees of autonomy from the private sector. We can distinguish five different analytical patterns of representation. This will be discussed in ascending order of separation and opposition to the system of private associations:

1 A first pattern is characterized by the fact that state-owned firms are represented by associations that do not distinguish membership according to the kind of ownership. In this case there is a good mediation between the two spheres of interests within the same associative system.

2 In the second pattern only the trade interests of public firms are represented by private associations. An example of this pattern is the British case, in which the state-owned firms directly join the CBI as a sector-unspecific trade association, but avoid being represented in the area of employers' interests, where the bargaining activity is carried out autonomously by the single public firms, according to the guidelines from the government bodies.[14]
3 The third pattern is a variant of the second, because it implies that interest representation is united only at the level of trade associations, while employers' interests are represented by two separate associative systems carrying out different industrial relations strategies, which are often in competition. A good example of this type is the Swedish case, where state-owned companies are members of the trade association of the chemical sector, but are also organized by the sector-unspecific employers' association of public firms.[15]
4 The fourth pattern shows a higher level of opposition between the two spheres. State-owned firms do not join either the employers' or the trade associations of private firms. They have their own employers' association, often in opposition to the private one, and their economic interests are directly represented by the relevant ministries, without organizing ad hoc associations. This is the case in Italy, where associations of public firms were created at the end of the 1950s by the major political party to break up the monopoly of business interest representation held by Confindustria. These associations also gave institutional recognition to the major trade unions in exchange for their acceptance of Western democratic political institutions.
5 The last pattern is characterized by the highest degree of separation and opposition between the two systems of interest representation. State-owned firms are members of specific associations in both functional areas, and associational policies are usually in opposition to their private counterparts. In this case the problems related to the logic of membership can be easily overcome by the associations of public firms through compulsory affiliation.

Looking at these five patterns we note that for both public and private firms it is easier to adopt cooperative behaviour in the field of trade rather than in that of employers' interests. The last pattern, moreover, is a very conflicting one, not evident in the cases examined above.

We need to remember that the relationship between the two

types of associations depends very much on the historical evolution of the overall relationship between public and private spheres in the economy. In the UK, for instance, the affiliation of public firms to CBI is relatively recent (1969), and is partly the result of successive waves of nationalization and privatization (Grant, 1986) produced by alternative Labour and Conservative governments. More specifically, the fact that the affiliation of state-owned firms to the CBI happened immediately after the renationalization pursued by the Labour government at the end of the 1960s, can be interpreted as a demonstration that the public sector is integrated into the private industrial world whose interests it shares.

Also in the Italian case, the chemical sector shows a shift from the sharp opposition between public and private spheres of the economy during the 1960s, to a progressive acceptance of state-owned firms in the Confindustria associational system, as a result of restructuring needs of the 1970s. This coalescence of interest became evident in 1984, when state-owned chemical firms became affiliated to the new association Federchimica, to ensure common representation of trade interests. Since 1984 the pattern of relationship changes from type (4) to type (3) in the chemical sector in Italy. In this country, as in the UK, the participation of public firms in the private business interest representation system has an image value, as they have been often accused of responding inadequately to crises because of their political conditioning.

Domain definition and boundaries

The associative domain of each association in the system can be defined by the number of firms that could become members according to the statutory rules. In practice during the research work it was necessary to distinguish between potential statutory domain and de facto membership, because the two spheres do not often coincide. For instance, in the Italian case the statute of Aschimici allows the affiliation of firms operating in the chemical sector and 'similar' sectors. ASAP affiliates firms belonging to the ENI group, no matter what sector they are in, and also firms operating in activities 'related' to the former. Unionchimica associates chemical and pharmaceutical firms without specifying size, even though it officially represents small and medium-sized firms. Therefore one can conclude that the statutes of the associations avoid precisely defining their domains, to allow maximum flexibility in their strategies of membership development and definition of associative boundaries.

The fact that associations prefer not to define their boundaries exactly can provoke a certain competitiveness among them. Cases of actual competition, however, are occasional and restricted to subsectoral areas.[16] In practice each associative domain is part of an integrated system where the fields of possible competition for the acquisition of membership are episodical.

The smallest associations usually represent an exception to the above rule, because the sharp specialization in interest representation can be useful in protecting more efficiently an homogeneous membership with limited resources. In this case the association seems to adopt a behaviour similar to that of a cartel. In fact the objective is to exclude external competition. Although it is not part of the sector under study, the case of trade associations in the machine tools sector in Italy can be cited as a classic example (Baglioni, 1983). If we turn back to our sector, the case of Unione Petrolifera, the Italian association of oil refiners, seems the closest to this mechanism. It is at the boundaries of the chemical sector as defined in our research, but its potential competition with an analogous association affiliated to Federchimica is tempered by a de facto division of membership by ownership: Unione Petrolifera represents foreign multinationals, while Federchimica represents home refiners.

The Italian case is an example of what has been called sectional overlap (Schmitter and Streeck, 1981): the main associations tend to enlarge their domain beyond the chemical sector, at the same time affiliating firms that belong only to certain chemical subsectors. In this way, unlike other countries, there is not a clear overlap between the associational domain and the sector, although Aschimici comes close. Even before it became a federation, this association has always been at the centre of a system in which some kind of competition for membership is based on the size of the firm and is met at subsectoral level. On the contrary, the division of the domain by ownership does not imply competition between Aschimici and ASAP, because public firms are clearly distinguishable from private ones. If we go into detail we find potential competition between Aschimici and Unionchimica, easily explained by noting the general competition between Confindustria and Confapi for the membership of small and medium-sized firms.

At the level of specific subsectors three associations could compete with the correspondent commodity groups within Aschimici. Surprisingly, the relations between each minor association and Aschimici are marked by cooperation. This can be explained by the increasingly influential position of Aschimici with respect to the others, a relationship that was evident when Federchimica was founded in 1984. In addition one notes that Federchimica did not change existing relationships between the associative system of the chemical sector and the principal association in the pharmaceutical sector (Farmindustria). Federchimica's and Farmindustria's affiliations with Confindustria facilitate their reciprocal coordination. Both associations jointly bargain collective agreements with unions. The boundaries of the respective memberships are clearly delimited, and representatives of one association sit on the governing body of the other. The logic of influence, more than the logic of membership, explains the lack of complete associational integration of the pharmaceutical sector into the chemical one. The relationship between firms and relevant ministries in the former imposes an ad hoc specialization of business interest representation and the creation of a capillary activity of assistance to the single firms, especially in their contacts with the ministry of health.

In the Swedish case we also find a dominant association that takes the form of a federation, and a number of minor independent associations that protect specific interests at subsectoral level. Sometimes these associations also represent interests of firms outside the chemical sector, as defined by the research project. This is the case of plastic materials associations that represent not only producers, but also transformers and consumers of plastic materials. A credible explanation of the structure of this domain is historical. The present situation is in fact the outcome of a period of competition between the associations in two neighbouring sectors, respectively KiK, the main chemical association, and the Federation of Engineering Manufacturers. Both are members of the peak sector-unspecific association, so that to avoid competition inside the system a new organization was established (SPF, Swedish Plastic Federation). SPF is not included in the overall associational system, but cooperates with KiK on an ongoing basis. The cooperation agreement includes the provision that firms may receive fee discounts if they join both associations. This clause was inserted in order to foster double affiliation.

Furthermore, this condition of coordinated autonomy allows SPF to mediate the interests of producers and consumers of plastic materials and also to affiliate institutional representatives, such as public research institutes engaged in plastic material standards and regulations, and the relevant departments of the Armed Forces. In sum, in this case the lack of overlap between associative domain and sector boundaries also depends greatly on the logic of influence.

On the other side we find a sharp functional partition of the interest representation system, which is certainly promoted, at least in part, by the fact that membership fees to the employers' associations are totally tax-deductible, while fees for trade associations are not.[17] This is why trade associations in particular have developed service companies for their associates. Selective incentives do not seem very important as far as employers' associations are concerned, due to the high level of centralization in the field of industrial relations. In fact, as we have already pointed out, collective bargaining is carried out by three associations that follow the division among private, public and cooperative firms. There is an employers' association at the level of the chemical sector only for the former type of firm, while for the others the sectoral specificity does not seem relevant, because industrial relations are managed at the level of sector-unspecific peak associations.

In the German case the functional differentiation of the interest representation system also makes a partition between employers' and trade associations. Nevertheless, unlike the Swedish case, a lower degree of centralization in industrial relations makes possible the constitution of a specific association in the chemical sector and therefore a sectoral specificity in the associational system is recognized.

The English case shows again how the plastic materials subsector tends to acquire associative autonomy from the chemical sector at large. Moreover we find a number of small organizations that can be defined as boundary associations. These are the result of the need for strong coordination between chemical producers and consumers in other sectors, as in the case of the Federation of Resin Formulators and Applicators (FERFA). FERFA includes producers of resins, pertaining to the chemical sector (as defined by the research project) and consumers, relating to the construction sector.

Selective and solidaristic goods as an incentive to membership

The role of selective goods as an incentive to associative action is stressed by the fact that the organizations under study usually supply a wide range of services exclusively for their members. The average proportion of the total organizational effort, in terms of staff commitment, expenditure and time spent on the provision of selective goods, is 25 per cent in the whole sector, a figure lower than construction, metal manufacturing and machine tools sectors, where the figure is 41 per cent. The importance of supply of selective goods tends to increase, the smaller the average size of firms in the associative domain. This relation is evidenced by a significant, although moderate, correlation index (-0.2015 $p = 0.007$).[18] Therefore the lower importance of selective goods in the chemical sector can be explained by the lower presence of small business, compared with other sectors. Small firms are in fact the main consumers of associative services, for they cannot afford an internal staff devoted to these functions.

In our sector the supply of selective goods is more concentrated on information activities for members, on the provision of technical publications and on the supply of economic, legal and fiscal counselling. The supply of selective goods in the field of industrial relations, in contrast, is relatively low in the associations of the sector. This depends on the level of centralization in industrial relations of each country, but can be explained also by the functional partition between employers' and trade associations, and by the fact that where industrial relations functions take place at central level the possible selective incentives are offered outside the specific interest representation of the sector. There are also a number of associations that offer individual services at market prices, regardless of whether firms are affiliated or not. Nevertheless these cases are not widespread and are sometimes associated with low membership fees, to overcome the problem of associative action.[19]

Summarizing the above, one can conclude that the chemical sector is not characterized by a particular development of selective goods, compared with other sectors under study. For instance, business interest associations in the machine tool sector supply a wider range of services to their affiliates, and the same is true in the construction sector.

As far as solidaristic goods are concerned, the findings of the international research project show that provision of this kind of associative output is marginal in the average of business interest organizations.[20] Specifically, associations in the chemical sector are less integrated than are other sectors in the production of this type of goods.[21] The importance of solidaristic goods is equally related to firm size. The smaller the firms in the associative domain, the more important is the type of goods in the total associational output.

Moreover, another important variable is the position of associations in the overall interest representation system. If the association is a first-level association, that is it directly affiliates firms or individuals, there are often a variety of activities and ad hoc initiatives devoted to socialization of members. In the case of second- or third-level associations, that is federations of associations, solidaristic goods are only produced indirectly, by promoting a positive image of the business community. Then businessmen are proud to identify themselves with it.[22]

It is also noteworthy that associations of individuals give more importance to solidaristic goods than do associations of firms. This is because the individual actor has more to gain from the creation of a collective identity than do firms, with their already established corporate identities. The association of firms is also more directly influenced by sectoral characteristics, limiting even further the development of collective identity. This is more easily found in associations with religious or political membership requirements, as in the Dutch case and in those countries in which we find the presence of cooperative firms.

Management of diversity

Business interest associations try to mediate between two organizational imperatives that are not completely compatible. On the one hand, they have to mediate and unify business interests according to the logic of influence. In fact the higher the percentage of representation in the sector the more likely the association will be able to influence the institutional stakeholders. The logic of 'number increasing' bases bargaining power on the number of affiliates in the association. On the other hand, maintaining a large membership base necessitates mediating a number of conflicting interests and representing them. This contradiction is at work in every sector, but takes on specific characteristics in the chemical area.

According to the theoretical suggestions of Schmitter and Streeck (1981: 146 and following), it is useful to distinguish between structural and functional parameters in the management of diversity. Structural parameters stem from the management of different interests that can arise from the fact that potential members occupy different positions in the market. Functional parameters pertain to the different interests among members that occupy the same structural position.

As far as the first type of differentiation is concerned, the chemical sector shows how the pyramidal structure of the associative systems can represent a variety of subsectoral interests through first-level associations that affiliate relatively small groups of firms sharing the same market positions.[23] Trade interests at subsectoral level, in particular, are better represented by small size associations. Their characteristic cohesion and specialization are also the result of delegating the representation of more general interests to second-level associations. In brief, the associative structure of the whole sector takes the form of a federation of

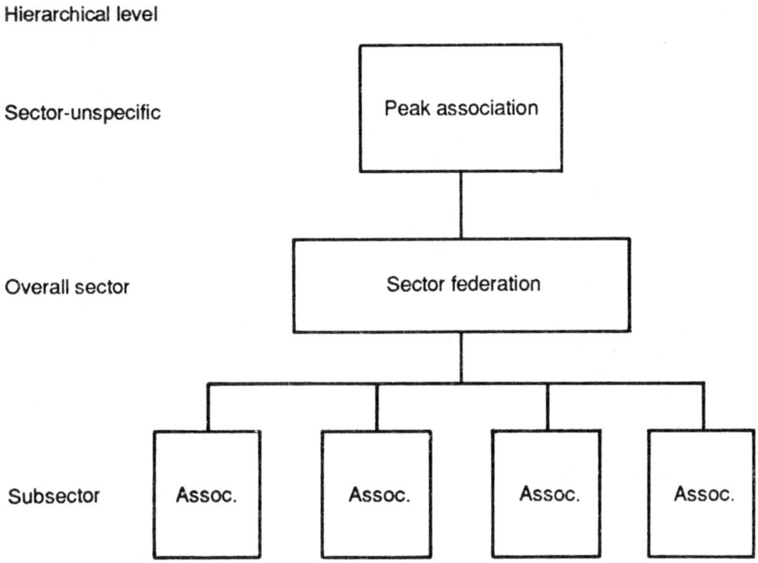

Figure 6.1 *The pyramidal model of chemical associative structure*

subsectoral associations and it is in turn affiliated to a sector-unspecific peak association which represents the industrial businessmen at country level. This structure is sketched in Figure 6.1, and can be functionally split in two separate systems as well. Of course the figure represents an ideal model, to which only the German case perfectly corresponds. Nevertheless, any deviation from this model can be understood by looking at the size of the sector in the country, the variety of its productions and/or other special conditions originated outside the sector, as discussed earlier in this chapter.

The Italian case, for instance, has developed very much along the line of this model since 1984. The Swedish case is partially different because the sector federation, affiliating ten subsectoral associations (Pestoff, 1983), is a mixed association and therefore many firms are directly affiliated with it. In other countries the chemical sector is not so large, because of the limited number of firms, and the associational system does not reach the above-mentioned scale and looks simplified and incomplete. The least articulated system is the Austrian one, because of the competing functions played by the chambers of commerce, as already discussed.

For these reasons we can roughly conclude that – after controlling for the external variables affecting business interest representation in the sector – the larger and more articulated the domain, the more likely a pyramidal structure, as depicted in Figure 6.1.

Nevertheless, the associative division between employers' and trade associations is the most important evidence of the need for interest articulation at the functional level. This partition can be explained by the fact that collective bargaining in industrial relations needs a unified strategy and practice. These are harder to obtain at the level of trade interests, because of the competition among firms and the potential conflicts of interest between producers and consumers of semi-finished goods within the sector. While this argument feasibly explains the clear functional division of the German representation system, it can be applied only partially to the Italian case, because it functions only in the representation system of state-owned firms and shows that it is more difficult to unify employers' interests than those of trade. Moreover, in the Dutch case it does not function at all.

Table 6.1 shows that there are seven times more trade associations in the sector than there are employers' associations. This is further evidence of the fact that the former have to take members' requirements more into account than the latter. Employers' associations usually enjoy higher bargaining efficacy if they are able to mediate and represent the whole sector before their counterpart. Thus the functional division between employers' and trade associations can be defined as an organizational device, maximizing the

Table 6.1 *Degree of functional specialization of business interest associations according to the sector and the country*

Category	Trade associations	Employers' associations	Both
Austria	15	0	3
Canada	34	1	4
Germany	17	7	12
Italy	2	3	11
Netherlands	23	2	21
Switzerland	25	5	11
United Kingdom	22	2	6
All			
Chemicals	30	1	2
Other sectors	73	12	43
Sector-unspecific	35	7	23
Total sample	138	20	68

unity of employers' interests. So it is possible to acquire a monopoly of representation, without denying an articulated representation of trade interests. This general rule has a stronger impact on the chemical sector than on other sectors, because of its clear-cut functional specialization, the limited number of employers' associations and the remarkable articulation of trade interests. This being the general rule, there are several exceptions at country level, related to the specificities of the larger systems of representation, of which the chemical sector is part. We can distinguish three types of exceptions:

1 Associative systems where the functional partition is less important. This is the case of the Netherlands and especially of Italy. A feasible explanation is found in the specific structure of the overall associational system. In Italy the dominant organizational pattern is that of Confindustria. Since the end of the Second World War important organizational changes affected the functional separation between employers' and trade associations previously coordinated only by a sector-unspecific peak confederation (Lanzalaco, 1988). In the Netherlands the functional partition was also abolished in 1968, as a result of the merger between the previous federations. Now other kinds of associative specializations are prevalent, such as a structure based on territorial criteria, which is in turn related to a religious dimension, while sectoral interests are represented inside this structure.
2 Associative systems based only on trade interest representation. This is the case of Austria, where a voluntary organization of business interests plays a residual role beside the compulsory organization of chambers of commerce, that have the monopoly of collective bargaining with the unions.
3 Systems where employers' associations are more numerous than trade associations. This is evidence of highly fragmented business interests in the field of industrial relations. This was again the case of Italy before the foundation of Federchimica.

Associations are able to utilize other mechanisms to unify interests and make their structure more congruent with the articulation of represented interests (Schmitter and Streeck, 1981: 27), but it does not seem that chemical sector associations need to refer very much to internal devices such as the ideological mobilization, the construction of specific collective identities or the resort to charismatic leaders. Rather, it makes use of differentiated voting rights to favour or penalize large versus small business. It takes advantage of the assistance and influence of the external representation system, of which it is a part,[24] as a way of improving the sector association's image before actual and potential members.

Another problem, emerging from an analysis of the organizational structure of the associations, deals with the choice between internalizing or externalizing membership diversity. This problem can be summarized by the following question: in what circumstances does a clearly recognized group of firms in a given domain develop into an ad hoc association, instead of a more or less autonomous group within a larger association? The problem can be confronted by taking into account the fact that the above is not a choice between two options, but rather a range of organizational solutions. In our sample we have found internal groups enjoying significant autonomy from the association they belong to; conversely, we have noted formally autonomous associations that are actually dependent on a higher level federation.

At least four factors affect this organizational choice:

1 This factor is based on the internal logic based on the relationship between organizational articulation and sector heterogeneity. This is the most general factor proposed by Schmitter and Streeck (1981) and is working *ceteris paribus*: a strong diversity of interests can be managed better within different associations than through a number of groups within one association.
2 A factor based on specifics of the associational environment. The looser the system the higher the likelihood that the diversity of interests gives origin to different associations.
3 A factor pertaining to the organizational history of the system. The present articulation is dependent on the previous one,[25] therefore the pre-existence of internal groups instead of autonomous associations can affect the organizational articulation of interest representation.
4 A factor relating to the interests of the organization as such. Business interest associations have their own interests to survive; to ensure prestige, power and incomes to their staff; to minimize changes. This factor is evident in the difficulties encountered by the longstanding attempt of Confindustria to develop second-order federations,[26] due to the vested interests of territorial sector-unspecific associations.

Finally, with the help of Table 6.2 we can sum up the main factors related to the logic of membership[27] that favour a simpler, more systematized[28] and integrated organizational structure in the chemical sector, than in other sectors analysed in the research project.

Table 6.2 *Average number of subunits of chemical associations compared with other relevant sectors*

Sector	Territory	Firm size	Task-specific committees	Group-specific committees	No. of members
Chemical	0.95	0.21	6.43	2.09	93
Pharmaceuticals	5.21	0.25	6.12	1.54	139
Construction	11.93	0.19	6.04	1.48	1254
Metal manufacture	9.47	0.06	6.47	2.12	1017
Machine tools	6.31	0.08	5.54	4.54	918

The first column of the table shows that the associative structure of the chemical sector is characterized by limited territorial articulation, because of the concentration of most plants in narrow areas and the integration of production in basic chemicals.[29] In fact the location of the chemical industry depends greatly on the state of technology in basic chemicals. That in turn implies large-scale economies, verticalization of production and spatial integration of the processes, because transportation of semi-finished goods affects prices greatly.

The last column of Table 6.2 shows that the chemical sector has the lowest average number of members for each association. As has been pointed out, the smaller the number of members the easier the conditions of collective action, because Olson's paradox (1965) tends to break down. Therefore we can conclude that the sector has relatively minor problems of interest representation compared with others quoted in the table.

On the other hand, the dominant position of a number of oligopolies in each country is a source of problems in the mechanisms of mediation between small and large businesses. For this reason associations in the chemical sector, as well as in pharmaceuticals, show a relatively large number of internal groups that explicitly distinguish and recognize specific interests according to size.

Finally, as I have tried to demonstrate in the preceding paragraphs, when the associative system in the sector is less integrated and systematized than the model shown in Figure 6.1, part of the explanation should be found in exogenous factors, that is more general features of the sector-unspecific system of business interest representation in which sectoral associations are embedded.

Notes

1. Therefore from a rational actor's point of view, free-riding ensures a clear advantage: the actor enjoys the collective benefit without paying for it.

2. The available historical studies agree on the role of these two interwoven factors based on differing situations in different countries, to explain the origins of business associations (Lanzalaco, 1988).

3. The second factor explains 18 per cent of the total variance between the nine items used during the interviews, while the proportion of variance explanation of the first factor is 23 per cent.

4. The firm's interest to become a member of a business association is different from that of businessmen and entrepreneurs to become involved in its representative bodies. This is because, unlike workers' representatives, businessmen can hardly use interest associations as a career opportunity. On the contrary, in most cases elected members have less time to devote to their business commitments.

5. In this chapter I shall use a database referring to the following countries: Austria, Canada, Germany, Italy, the Netherlands, Sweden, Switzerland and the United Kingdom. For each country I have made comparisons between chemical sector and construction, metal manufacturing and machine tools sectors. Pharmaceuticals have been considered separately, because in some countries they are part of the chemical interest representation system, in others they are not. Although the data were collected also for textile and food sectors, I did not include them into the international comparison, because they are not available for all the above-mentioned countries.

6. According to the data gathered by the research team, the number of firms circa 1980 ranged from 106 in the Netherlands to 4169 in Italy, with intermediate cases such as Great Britain (755) and Sweden (315).

7. In Great Britain ICI accounts for 37 per cent of the total output in the sector. In the Netherlands the state-owned DSM accounts for about 25 per cent of the total chemical industry. A similar proportion is found in Italy for Montedison. The three main oligopolies in West Germany supply more than 50 per cent of the production, while in Austria the figures are 28 per cent for the largest firm and 52 per cent for the first four.

8. Although it is an example pertaining to a sector outside the one we are now examining, as evidence of this mechanism we can quote the 1983 contract renewal in the textile sector in Italy (Guidotti, 1986). A more general analysis of the associative system in the textile sector is given by Mutti et al. (1986).

9. The CBI has been defined by Grant an 'umbrella association' rather than a peak association of an integrated system.

10. Trade union membership reached 90 per cent of total employment during 1970s.

11. A mixed association has been defined by the research project coordinated by Schmitter and Streeck (1981), as an organization that affiliates both single firms and associations of firms. In this case, beside the single-member firms, VNCI affiliates four subsectoral associations.

12. This role is also related to three factors: the end of the strong oligopolistic competition, that took place as a consequence of the energy shock; the definition of the reciprocal sphere of influence between the public and private capital in the sector; and the legitimation role played by Confindustria, the peak association of the business system, in favour of the constitution of Federchimica (Chiesi, 1985).

13. Until the recent affiliation to Federchimica the state-owned firms in the sector have never been represented by a trade association. This can be explained by the fact that their owner is the Ministry of Industry itself. According to the associational staff this overlapping between the shareholder and the relevant state agencies has avoided the need for lobbying.

14. According to different authors the government policy exerts a considerable influence on industrial relations in public enterprises. This is the case of United Kingdom (Grant, 1986), and Italy (Martinelli, 1981).

15. A similar system of associative relationships is found in Sweden not only for state-owned firms, but also for the interest representation of cooperative firms (Pestoff, 1983).

16. Different authors agree on this evidence (Grant, 1983; Chiesi, 1985; Traxler, 1983).

17. The importance of the institutional context and the actual legislation of each country, in influencing the organizational structure of business interest associations, has already been pointed out by Schmitter and Streeck (1983).

18. The relation between dimension of the firm and interests for the provision of selective goods has already been pointed out in a number of studies (Grant, 1986; Martinelli and Chiesi, 1988).

19. Although the research project has collected a lot of data on this issue, the problem of efficacy of the different mechanisms of fund-raising for voluntary associations is going to remain a matter of evaluation.

20. In part this marginal position is due in my opinion to the fact that the data collection criteria for each association are actually limited to the official production of services apparently related to the provision of solidaristic goods such as fund-raising for charity, members' participation in the general assembly and the organization of social activities for members. The research project does not take into account a considerable variety of associative activities that have a double function and therefore also play a role in promoting collective identity. In this respect we can recall conferences and seminars among members, and also the frequent opportunities of meeting and exchange of opinions among them that associations constantly promote. We can therefore state that the scanty visibility of solidaristic goods provided by the associations is in part due to the fact that they are sometimes the unintended consequences, or a byproduct, of a variety of associational activities.

21. The variable, which has been used to measure the supply of solidaristic goods in terms of organizational effort, is computed summing up the scores ranging from 0 to 2, for each activity, according to the importance attributed by the associational staff.

22. Of course, a good image of the business world is primarily a public good. Martinelli and Chiesi (1989) have shown how the promotion of a good image, related to the development of a managerial culture, which is shared by public opinion, is a relatively new issue in Italy, and has contributed to strengthening the prestige of oligopolies during the 1980s. But here we make reference to the image of the associational system in itself (that is, to be member of Confindustria), and to its consequences on collective identity.

23. As described in preceding paragraphs, the subsectors of pharmaceuticals and plastic materials are the most likely to be organized in specific associations in most of the analysed countries. Some basic and fine chemical productions are sometimes organized in specific associations, or give origin to product groups inside a larger

association, and are endowed with different degrees of autonomy.

24. See for instance the role played by Confindustria, the peak association of the industrial system of interest representation, in giving prestige and legitimation to Aschimici.

25. This dependence has been thoroughly analysed by Lanzalaco (1988) at the level of peak associations in different countries.

26. Although Federchimica is a successful example of this attempt.

27. Here I omit to quote the explaining factors pertaining to the logic of influence, as they are discussed in Chapter 7.

28. The importance of evaluating the degree of systematization in the analysis of an associative system has been pointed out by Lanzalaco (1988).

29. The sectoral interdependence of firms can reach considerable values, as in the UK, where 70 per cent of the total chemical sales are processed within the sector before being exported or absorbed by the retail market (Grant, 1983).

References

Baglioni, M. (1983) *L'azione collettiva*. Milan: Giuffrè.

Baglioni, M., Chiesi, A.M., Maraffi, M. and Martinelli, A. (1983) 'L'organizzazione degli interessi imprenditoriali in Italia: primi risultati'. Paper presented at the third 'Organization of Business Interests' Project Workshop, Florence, European University Institute, 19–21 May.

Chandler, A.D. (1977) *The Visible Hand*. Cambridge, MA: Harvard University Press.

Chiesi, A.M. (1985) 'L'organizzazione degli interessi imprenditoriali in un caso di ristrutturazione settoriale'. *Studi Organizzativi*, no. 3/4.

Chiesi, A.M. and Martinelli, A. (1988) 'L'organizzazione degli interessi imprenditoriali come meccanismo di regolazione sociale', in P. Lange and M. Regini (eds), *State, Market and Social Regulation: New Perspectives*. Cambridge: Cambridge University Press. pp. 187–213.

Coleman, W.D. (1985) 'Quebec nationalism and the organization of business interests'. Paper presented at the conference 'The Regional Organization of Business Interests and Public Policy', McMaster University, Hamilton, 22–24 May.

Grant, W. (1983) 'The organization of business interests in the UK chemical industry'. Paper prepared for the 'International Conference on Business Interests Associations in the Chemical Sector', ASAP, Rome, April.

Grant, W. (1986) 'The politics of private business and public enterprises in the United Kingdom'. Florence, European University Institute, Colloquium Papers.

Guidotti, D. (1986) *Strategia generale e azione decentrata*. Milan: F. Angeli.

Kriesi, H.P. and Farago, P. (1985) 'The regional differentiation of business interests associations in Switzerland: economic, political and social determinants'. Paper presented at the conference 'The Regional Organization of Business Interests and Public Policy', McMaster University, Hamilton, 22–24 May.

Lange, P. and Regini, M. (eds) (1988) *State, Market and Social Regulation: New Perspectives*. Cambridge: Cambridge University Press.

Lanzalaco, L. (1988) 'La rappresentanza degli interessi a livello territoriale: il caso delle associazioni imprenditoriali italiane in prospettiva comparata'. PhD thesis,

European University Institute, Florence.
Martinelli, A. (1981) 'Confindustria e potere politico', in A. Martinelli, A.M. Chiesi and N. Della Chiesa (eds), *I grandi imprenditori italiani*. Milan: Feltrinelli. pp. 235–82.
Martinelli, A. and Chiesi, A.M. (1988) *La rappresentanza associativa imprenditoriale di fronte al cambiamento degli anni '90*. Milan: Istituto Superiore di Sociologia.
Martinelli, A. and Chiesi, A.M. (1989) 'The capitalist class in Italy', in T. Bottomore and R.J. Brym (eds), *The Capitalist Class: An International Study*. Brighton: Wheatsheaf.
Martinelli, A., Chiesi, A.M. and Della Chiesa, N. (1981) *I grandi imprenditori italiani*. Milan: Feltrinelli.
Martinelli, A., Schmitter, P. and Streeck, W. (1981) 'L'organizzazione degli interessi imprenditoriali', *Stato e Mercato*, no. 3.
Mutti, A., Addario, N. and Segatti, P. (1986) 'The organization of business interests: the case of the Italian textile and clothing industry', Working Paper no. 86/205, European University Institute, Florence.
Olson, M. (1965) *The Logic of Collective Action*. Cambridge, MA: Harvard University Press.
Pestoff, V. (1983) 'The organization of business interests in the Swedish chemical industry'. Paper prepared for the international conference 'Business Interests Associations in the Chemical Sector', ASAP, Rome, April.
Pizzorno, A. (1983) 'Sulla razionalità della scelta democratica', *Stato e Mercato*, no. 7.
Ranci, P. and Vaccà, S. (1979) *L'industria petrolchimica in Italia: anatomia di una crisi*. Milan: F. Angeli.
Sargent, J. (1983) 'The politics of the pharmaceutical price regulation scheme'. IIM Discussion Paper 83-13. Berlin: Wissenschaftszentrum.
Schmitter, P. and Lanzalaco, L. (1988) 'L'organizzazione degli interessi imprenditoriali a livello regionale', *Stato e Mercato*, no. 22.
Schmitter, P. and Streeck, W. (1981) *The Organization of Business Interests: a Research Design to Study the Associative Action of Business in Advanced Industrial Societies of Western Europe*. Berlin: Wissenschaftszentrum.
Schmitter, P. and Streeck, W. (1983) '"The rules of the game" for business interest associations: an addendum to the research design on "The Organization of Business Interests"'. Paper presented at the third 'Organization of Business Interests' Project Workshop, Florence, European University Institute, 19–21 May.
Skogh, G. (1984) 'Employers' associations in Sweden', in J.P. Windmuller and A. Gladstone (eds), *Employers' Associations and Industrial Relations*. Oxford: Oxford University Press. pp. 149–68.
Solé, C. (1985) 'The system of organized business interests in the Spanish textile industry'. Paper presented at the conference 'The Regional Organization of Business Interests and Public Policy', McMaster University, Hamilton, 22–24 May.
Streeck, W. (1986) *The Territorial Organization of Interests and the Logics of Associative Action: The Case of Artisanal Interests Organization in West Germany*. Berlin: Wissenschaftszentrum.
Traxler, F. (1983) 'Structure and functions of the business interests associations in the Austrian chemical sector'. Paper presented at the international conference

'Business Interests Associations in the Chemical Sector'. Rome: ASAP.

van Woorden, W. (1984) 'Employers' associations in the Netherlands', in J.P. Windmuller and A. Gladstone (eds), *Employers' Associations and Industrial Relations*. Oxford: Oxford University Press. pp. 202–31.

7

The Functions of Associations as Agents of Public Policy

Henry J. Jacek

This chapter deals with the industrial problems facing chemical manufacturers, and raises the question whether strong industry associations can provide the leadership in solving some of these problems. The concept of relative autonomy (*Verselbständigung*) of the business interest association (BIA), which refers to an organization's ability to control its supply of resources and to its capacity to select its goals and the strategies and tactics it believes necessary to achieve those objectives, is the critical property of strong interest associations. As the autonomy of interest associations increases, then these organizations could become private governments[1] or more appropriately 'private interest governments' as defined by Streeck and Schmitter (1985: 14–17). These associations are seen as being given public responsibilities so as to become organizations of 'regulated self-regulation. In this way special interests . . . are made subservient to general interests by appropriately designed institutions.' The 'collective self-interest' of these organizations is used to generate and reproduce a stable social/economic/political order because it is assumed that these associations are capable of being 'transforming agents of individual interests' so as to produce 'responsible associative governance'. So in this chapter 'private interest government' will be used as 'the concept for arrangements under which an attempt is made to make associative, self-interested collective action contribute to the achievement of public policy objectives'. In this way chemical industry associations may function as agents of public policy and help solve sectoral problems.

The range and character of possible private interest government functions exist in at least five major policy areas[2] that affect the chemical industry; industry structure including overcapacity, labour, supplier/customer relations, standards, and general macro/intersectoral economic policies. The first major purpose of this chapter is to order these five policy areas along a continuum which distinguishes these areas in terms of the attempts to solve industry problems by means of collective solutions. This exercise hopefully is an elaboration of Streeck and Schmitter's (1985: 28) generalization that 'there is growing evidence that there is a certain range of policy areas for which institutions of group self-regulation may produce more socially adjusted and normatively acceptable results than either communal self-help, free trade, or *étatisme*'.

The relationship between the organizational development of business interest associations, especially as indicated by the concept of relative autonomy, and private interest government is based on the view that the successful assumption of private interest government functions can lead to the provision of more and more resources by the state and the member firms. Such provision in turn is likely to lead to greater strategic autonomy both from the association's individual member firms and from the state, the public government. The functions of private interest government are thus both an indicator and outcome of organizational autonomy.

The second important part of the chapter will examine the extent to which the connections exist between resources components and private interest government outputs. First on the input side questions can be raised about financial resources and public status. Do chemical industry associations that act as private interest governments have a more diversified financial resource base compared to those associations that lack governing properties? Sources of finance and the relative importance of independent professional staff capacity of associations versus voluntary members' assistance are the important indicators here. Also, to what extent is public status[3] attributed to private interest governments vis-à-vis poorly developed associations? The former are more likely to be formal participants in para-state bodies. In addition these same organizations are expected to be holders of state-granted licences to administer particular goods.

Second, do private interest governments provide monopoly goods to a significant extent? Monopoly goods are goods or services to which an association has been granted sole or monopoly rights in their provision (Schmitter and Streeck, 1981: 229–38). What other types of goods are produced by BIAs and what is the balance between these types and monopoly goods? Also, to what extent is there any pattern of provision of goods by different types of subsectors, countries and associational structures? In particular, is the provision of monopoly goods subject to explanation by these other factors?

Beyond the very important question of the possible contribution of private interest governments to solving industrial problems, in this case the problems of the chemical industry, lies a very basic theoretical contradiction in the public policy writings by contemporary social scientists. The unresolved question is the relative importance of national styles of policy-making versus the properties of policy areas in producing the conditions that are conducive to collective yet private solutions to industrial problems. Are business interest associations acting as private interest governments explained better by conditions specific to certain countries, or by the ability of private interest governments to handle certain kinds of problems that arise in particular policy areas?

The received wisdom of policy analysts is that the history of national cultures and institutions predicts the emergence of private interest government and the more general corporatist solutions to economic problems. The organized complexity of the institutions or organizations of political economy and their use in mastering economic challenges are contingent, so it is believed, on the historically determined nature, whether centralized or fragmented, of these institutions. This argument also emphasizes the national political interrelationships or the underlying balance of forces among business, labour, the state and the various political parties as a key determinant for the successful cooperation among these organizations in governing the national political economy. Thus, the possibility of private interest government as a mechanism and a viable answer to the oscillations of economic opportunities and disturbances is *contingent* on national histories.

Such an approach underlies a great deal of work in comparative public policy. Recently dramatic counterintuitive differences in the regulation of occupational health and safety in the chemical industries of the United States and Sweden were uncovered using the notion of national policy style (Kelman, 1981). A number of works in the Cornell Studies in Political Economy series analyse the comparative scene from this perspective. Peter J. Katzenstein (1984, 1985), the series editor, places an emphasis on the size of political economies in the world system as the basic influence on the relevant nation. In the same series, David Vogel (1986), in his analysis of environmental policy and the chemical industries in Britain and the United States, develops the concept of 'national styles' as 'rooted in the history of its pattern of interaction among business, government, and the public' (p. 264). This pattern in turn is based on two foundations: the constitutional regime and the political values of the population.

While we would not deny the importance of national variations in business–state historical relations, is this the whole story? Some policy theorists think not. The contrasting view is that certain policy areas are more conducive to private interest government solutions than other areas. This specific argument is part of the more general view that different statutory policy instruments and various organizational policy agents work most effectively when specific mixes of instruments and agents are applied to problems in the appropriate policy area. Despite six chapters on six different countries Dyson and Wilks (1983) work from this perspective. Also explicit in their analysis is the view that countries can learn from one another – a distinct advantage of comparative public policy even in such a seemingly intractable area as 'industrial crisis' and despite the predisposition of national histories to certain solutions. 'Nevertheless, comparison offers to policymakers interesting insights on which to base institutional and programmatic reforms' (1983: 261).

Types of public policy and the functions of business associations

The five major policy areas studied in this chapter may be divided into policy subareas. Greater elaboration of the form of private interest government for these policy subareas will be provided. Analytically, the policy subareas are outlined below.

The shaping of industry structure may be divided into four policy subareas: investment and industrial restructuring, competition, research and development, and prices and profits. The coordination of investment and deinvestment, that is rationalization of the industry, if done privately, could be accomplished by joint ventures, mergers, reorganization of particular firms including changes in their pattern of ownership (Grant, no date), or possibly by cartels, but if there is public involvement BIAs may administer government funds or may promote larger units of production. The second subpolicy area, which is possibly an association activity where such behaviour is not illegal, is shaping industry structure by regulating competition among firms by forming input prices and the allocation of raw material supplies, appropriate market shares and the development of other competitive or anti-competitive practices such as measures that prevent the entry of new competitors. Of course, these are ways for protecting or preserving the status quo, or at least for controlling the pace or direction of change. A third policy subarea that may affect industry structure is research and development. If research and development is solely an activity of the firm then firms can be distinguished from one another by the quality of research and development (R&D) if it is done at all. Firms with high-quality R&D can be expected over time to increase sales, assets and profits at the expense of competitors, leading to tendencies towards oligopoly. If, however, a BIA carries out industry research and development then firms, probably small ones, with an individual competitive disadvantage potentially can have access to industry state-of-the-art research and development. The final policy subarea of the general area of policies that affect industry structure is the one involving policies on output or prices. In the chemical industry these prices are distinctive in that a large proportion of sales are either within companies, and thus not real market prices or costs, or to other firms in the industry. These sales may be done on a long-term stable foundation which are negotiated contract prices or on a short-term more volatile basis which are made on the spot subject to immediate market forces. These policies may be the most important because they seem most likely to affect the very profitability of the industry. One might expect capitalism unfettered by either the state or association to be the most profitable.

The second broad category of public functions that chemical industry associations may perform are policies that regulate the industry's labour market. Possible policy subareas include negotiating binding labour agreements, coordinating practices on the hiring and firing of workers, administering state vocational education, and administering worker social insurance programmes. The usual type of labour regulation that first comes to mind is the negotiation of industry-wide, multi-firm labour agreements including binding agreements on minimum wages and conditions, in general the promotion of labour peace. Related to the first policy subarea are policies that involve the hiring and firing of workers. Policies in this subarea may have as goals labour blacklisting or attempts to find sources of low-paid guest workers. Third, an association either by itself or in cooperation with state agencies may administer state vocational education and training. The final subarea of policies affecting labour is in the administration of worker social insurance programmes including pension plans.

The third general type of public policies are those that help stabilize the chain of production relations from upstream commodity suppliers to downstream transporters, distributors, domestic retailers and agents for foreign sales. Five policy subtypes will be examined in this section: the *recognized right* (Offe, 1981: 135) to negotiate or participate in decisions on prices, quotas, supplies and allocation of supplies of raw materials with commodity producers and firms covered by their domain and with retailers, wholesalers or buying groups; the promotion of the use of domestic inputs;[4] commercial arbitration; development of industry markets including exports; and the collection of authoritative information from member firms for use by state agencies in the making of public policy. The first subarea could cover the negotiation of binding agreements with all customer or supplier organizations on prices and other conditions. Second, a business interest association may, on the urging of the state, promote the use of domestic inputs in place of foreign supplies. For example a chemical industry association may promote the use of domestic commodities such as feedstocks among its member companies even though these may be more expensive than offshore supplies. A third subarea is commercial arbitration and in a fourth subarea, BIAs may promote the marketing, including exports, of the industry's products. The fifth and final subarea of the 'stabilization of the chain of production relations' general area is the collection of authoritative information from member firms for use of state agencies in the making of public policy.

The fourth broad policy area that may be taken over by business interest associations in the chemical industry is to regulate quality and health standards, which in general could be seen as protecting the public interest of consumers and society at large.

The following subareas will be examined below: product quality; hygienic manufacture and safe handling of chemicals; standardization of products; labelling; control of the use of labels and hallmarks; the publication of standards and expectations of the industry's products; standardization of business forms and contracts and industry regulation of advertising. The first subarea deals with product quality. A business interest association could be charged with developing, promoting and most importantly enforcing quality standards, grades and standards of conduct. Second, there is the hygienic manufacture and safe handling of chemicals, for example, defining a proper standard of handling toxic chemicals or the procedures for cleaning up chemical spills. This would possibly involve training of workers in the proper handling of toxic chemicals. A third policy subarea is the standardization of products by both qualities and quantities available to consumers. Fourth, an industry association can promote the proper listing of ingredients and their relative proportions in such things as domestic chemical specialties, for example garden pesticides, as a way of overcoming a typical market failure. A fifth and very important set of policies is the control and use of labels and hallmarks. Sixth, quality and health standards may be promoted by the joint association/state publication of standards and expectations, and the seventh policy subarea is the standardization of business forms and contracts followed by the final subarea of industry regulation of advertising.

The fifth and final set of broad public policies are those that are not specific to one particular industrial subsector but involve broad public policy goals for the chemical industry as a whole, for domestic economic policy or for international policies. In this area BIAs are used as agents to help the state implement its broadest policy goals. We can identify several ways by which BIAs can be agents of public policy: encouraging compliance with state laws in general and thus generating diffuse legitimacy for the state; by routine consultation in drafting state legislation and regulations; helping the state to reach its macroeconomic goals; helping to implement state regional policy; involvement in environmental pollution control; and participation in formulating their government's position on the development of supranational business policies. The first subtype involves activities that encourage the general compliance with not only the letter but also the spirit of state laws and regulations. This process involves continual education and encouragement of member firms to follow state policies as a matter of course. The association staff may translate public policies from state bureaucratic language into the common language of business. A second specific type of association activity in this broad area of intermediate or macro-level state economic policy is the routinized and recognized right to be consulted in drafting state legislation and regulations. This right is important in helping the state implement its public policies by improving compliance because such formal consultation, together with agreement or compromises before a government gets committed to laws and regulations, is expected to lead to business compliance once the public policy is proclaimed by the state. Third, business interest associations may be mobilized to help the state reach its macroeconomic goals such as growth, productivity and the restraint of inflation, and to create high levels of employment. Fourth is the state's use of associations to implement state regional policy, and a fifth general policy subarea is state energy policy goals. A sixth possible area is to help the state implement state environmental policy. A final public policy subarea would be to help the state in the formulation of its position on supranational chemical industry policies such as the *Codex Alimentarius*.

Policy areas and their public character in Europe

After a careful examination of the amount and form of private interest government in a number of countries one conclusion is clear, namely, private interest government is found in all countries including those normally thought of as pluralist (Jacek, 1987) or as having a significant traditional, informal non-modern sector of the economy (Berger and Piore, 1980). When interest associations act as agents of public policy they do not seem to be inhibitors of economic growth. Rather, by being closer to chemical firms than are government officials they seem able where legally possible to move quickly to solve common industry-wide problems once they arise, whether within a country or on a multinational regional or global basis.

Austria's chemical industry provides us with the most developed system of associational control. For the most part private interest government occurs through a unique relationship between the voluntary interest associations which represent the business interests of the chemical sector and the legally compulsory industrial chambers of commerce (*Kammerorganisation*) on the one hand, and between the chamber organization and the state on the other. The normal situation is where the chamber organization is given de facto recognition in policy-making and implementation but this organization lacks the technical expertise to act effectively alone. In this situation the chamber organization does the actual negotiating but with the advice and help of the voluntary associations. It is the voluntary association which prepares the chamber organization for price and wage negotiations. The argumentation is prepared for the formal negotiators who lack the necessary technical and medical knowledge.

This relationship occurs between the same chamber organization that represents drug, medicine and chemical, aerosol manufacturing and the voluntary association of manufacturers of chemical aerosols. The latter are represented by the Austrian Aerosol Association (Österreichische Aerosol Vereinigung – ÖAV). For aerosols there are many safety standards affecting their production, storage and transport. When these standards are negotiated with the state by the chamber organization, the latter is given the necessary technical information by the ÖAV (Traxler, 1983).

This specific example of Austrian private interest government takes place within the far more important centralized institutional complexity of Austrian interest politics (Marin, 1983, 1985). In addition to voluntary sectoral business associations such as the ÖAV there exists the voluntary League of Austrian Industrialists, the Vereinigung Österreichischer Industrieller (VÖI). These voluntary associations complement the more widely known compulsory chambers by giving special representation to large firms or to sectoral and subsectoral interests with distinctive properties or problems. The Swedish system is centralized, but in a different way from the Austrian. The labour policy area provides the typical example. The Swedish chemical industry has three employer associations, Almana Gruppen Svenska arbetsgivarefoeningen (Ag-SAF or the General Group of the Swedish Employers' Confederation), KFO (the Cooperative Labour Negotiation Organization) and SFO (the Employers' Association for Companies in Joint Ownership). The Ag-SAF organizes companies from the chemicals, drugs and chemical engineering industries, and also minor branches of industry, mining and various service trades. It is the third largest of the thirty-six branch associations that belong to the SAF. According to its statutes, branch associations and their member firms have a duty to follow SAF statutes and the organizational decisions. If they do not fulfil their duty, fines may be levied by SAF's General Council with no appeal possible. If a lockout is called, member firms that are not affected are not allowed to 'provide employment for workers affected by a lockout, or in any other way support them; the same applies to workers who go on strike'. Fines are also levied for companies breaking collective agreements and thereby causing a wage drift (Bengtsson et al., 1984: 25 and 58).

The KFO's member firms are mostly concentrated in the food-processing industry, but it negotiates for a number of chemical companies. The KFO reserves to itself the right to agree to an industrial relations agreement for its member firms. Should a labour dispute occur, the member firms must follow policy directives that come from KFO. If these member firms do not abide by KFO regulations they are liable to expulsion. Members of SFO must also follow its statutes; otherwise they forfeit all claims to SFO's strike fund. The governing board of the SFO 'can decide upon a fine or the expulsion if a member willingly circumvents the statutes or a decision in agreement with the statutes, or if it fails to fulfill its responsibilities according to the statutes' (Pestoff, 1983: 25).

In the chemical industry an important labour matter is the occupational safety and health of its workers. In Sweden these matters are safeguarded by the *Arbetsskyddsstyrelsen* (ASS – the National Board of Occupational Safety and Health). Of the eleven persons on this Board, employers nominate two representatives, one from SAF itself and one from a branch organization, the Engineering Employers' Association.

In addition these employer associations administer large government grants for various projects. All three employer organizations 'run courses on co-determination legislation and its implementation. These courses are financed by the ASF' (the National

Occupation Safety Council). Other courses related to the workplace are sponsored by both KFO and SFO and also financed by ASF. The SO (the National Board of Education) also 'finances the joint activities of all three' employer organizations in running occupational training courses. Finally, one trade association, the KIK (the Association of Swedish Chemical Industries) received state financing for worker education in the handling of hazardous goods and chemical substances (Pestoff, 1983: 69).

In Sweden KIK represents not only the industrial chemical industry but also the drug manufacturers. The National Board for Health and Welfare's Department of Drugs has a responsibility for controlling drugs. Attached to this Department is the Board of Drugs, which is responsible for deciding which drugs will be registered for sale in Sweden, and which drugs will be removed from both human and non-human consumption. The SI (the Federation of Swedish Industry) appoints the Director General of KIK as its representative on this Board (Pestoff, 1983: 59).

There are numerous examples in which Swedish chemical associations act as agents of public policy in the area of standards. A striking example is the 1974 framing of 'regulations concerning maximum permissible exposures to chemicals'. Essentially these far-reaching standards were devised by three people: a chemical engineer who was the head of ASS's Office of Standards Department, a SAF health expert and a representative of the Landsorganisationen (LO – the Swedish Trade Union Confederation). In some cases there was a drastic revision downward in acceptable limits of exposure. The SAF representative approved these lower limits without any consultation with individual firms. Yet some of these limits were so low that some chemical plants were forced to close down because of the drastic revision (Kelman, 1981: 62–70).

The Produktkontrollnenden (PKN – the Products Control Board of the Environment Protection Board) was created in 1973 to set standards on product control involving potentially hazardous products. The chairman of KIK sits on the PKN. This Board registers pesticides, issues permits for the import and sale of poisons, checks that hazardous products are labelled with proper warnings and instructions for safe use, and requires that it be provided with adequate information about these hazardous products (Pestoff, 1983: 53–4).

Finally in the area of standards the SPF (the Swedish Plastic Federation) nominates four members of the Committee of Standards for Plastics of the Swedish Standardization Commission. The central objective of the SPF seems to be standardization. It is concerned with the establishment of functional and material norms for plastic products, standard approval certificates and 'standardization of quality and size of various products in conjunction with the Swedish Standardization Commission' (Pestoff, 1983: 32–4, 68).

The associations of the Swedish chemical industry are made part of the process of devising and implementing macroeconomic policies through the use of ad hoc parliamentary Royal Commissions and the process of remitting their reports to the association, for criticisms and revision. This system is analogous to the Austrian Partitäische Kommission. The Swedish associations are an integral part of these Royal Commissions and the *remiss* process which involves written briefs in response to the Royal Commission reports and the Commissions' study and analysis of the *remiss* briefs (Pestoff, 1983: 46, 66–73).

Of all the macroeconomic policy areas within which chemical BIAs can act as agents of public policy, none seem to be more prominent than environmental policy. The Statens naturordsverk (SNV – the National Environment Protection Board) was started in 1967 as the central administrative authority in the environmental policy area. The SI (the Federation of Swedish Industry) nominates one member to serve on the SNV Board of Directors as well as two of its advisory bodies, the Water Protection Council and the Air Protection Council. As well there is the Konsessionsnämd förmiljöskydd (KNMS – the Franchise Board for Environment Protection) which considers applications to carry on polluting activities. This board has four members, one of whom is nominated by SI (Pestoff, 1983: 51–5).

When associations in the remaining countries take on the characteristics of private interest governments it is not on a centralized multisectoral basis but on a more sectoral or subsectoral basis. However, this is not to say that the presence of private interest government is minor in these cases, rather it is the form that is different. Britain is a case in point where there is quite a lot of private interest government by British BIAs in the chemical industry. While this may seem surprising to some, an important element of contemporary politics in Britain is the reduction of state bureaucracy and the transferring of policy implementation to interest associations (Grant, 1984).

Although the British pharmaceutical subsector, an industry closely related to industrial chemicals, has had high profits in good times and bad, the industrial chemicals industry has been badly hit by recessions and severe overcapacity. Because of this action has been taken in the plastics subsector. Here the British Plastics Federation (BPF) 'approached the Department of Industry for help with a comprehensive rationalization scheme in the processing sector. The Industry Department has stated that specific proposals from the plastics industry would be welcome and these are being drawn up by a senior working party of BPF members'. The leadership inside the BPF believes that the need for rationalization in the plastics industry must be triggered by outside support (Grant, 1983: 41–3). British BIAs, and especially the major one, the Chemical Industries Association (CIA), are organized to take a long-term strategic view of the entire sector. Partially this is a result of the impact of the EEC, which has forced the CIA 'to take a broad national view of its domain and to develop some sense of what the longer term holds' (Coleman and Grant, 1984: 219).

It is in the area of labour policy, especially worker training, that we can see the clear devolution of state functions to industry controlled bodies. 'The Chemical Industries Association has assumed full responsibility, in conjunction with the leading associations in pharmaceuticals, for developing and administering a programme of occupational training in the British chemical industry. This is a major role and one that will have long-range consequences for the sector' (Coleman and Grant, 1984: 220). This association 'has established a new Training Department . . . in its Industrial Relations Directorate'. This department, together with other chemical BIAs, has set up the Chemical and Allied Industries Training Review Council (CAITREC), which monitors and encourages training activities by other bodies (Grant, 1983: 64).

Even before the present period the British government used employer associations in the chemical industry as agents of public policy. In both world wars pressure was put on individual chemical companies to join chemical employer associations for the purpose of engaging in tripartite negotiations. Here the state used its power as a major customer, something seen over and over again in the British chemical industry. Sanctions were imposed on those firms resisting state pressure. The motives of the state were clear. First, there was a desire to reduce or eliminate disputes in industrial relations. Second, the British state as a major customer wanted predictable labour costs (Gill et al., 1978: 47).

British chemical associations are active in customer–supplier relations with the backing of the state. In the autumn of 1982 the British Plastics Federation started a Plastics Advisory Service. This new service received £120,000 in state aid from the Department of Industry. Help is given in three areas: (1) to suppliers of materials products, machinery and services; (2) on the 'technical aspects of product design, material selection and manufacturing techniques'; and (3) on 'commercial aspects such as market information and statistics on an international scale' (Grant, 1983: 50–1, 66).

There are numerous cases of self-regulation in the policy area of standards. The British Agrochemicals Association (BAA) requires all its members to conform 'with the provisions of the Pesticides Safety Precautions Scheme', which is approved by the state's Office of Fair Trading. This scheme requires that manufacturers do not sell pesticide to wholesalers not approved by the British Agrochemical Supply Industry Scheme (BASIS), which itself has state sanction (Grant, 1983: 29–30).

Advertising standards are the most frequent case of association self-regulation in standards, however. In the BAA's code of advertising there is a detailed set of industry standards. One puts limits on the use of statistics as well as limiting the types of things that can be said about competitors' products. The purpose of these advertising regulations is to prevent the subsector from falling into disrepute, a measure that is seen as necessary even though the private property rights of pesticide firms are being infringed (Grant, 1983: 32–3).

Perhaps the most decisive instance of private interest government involved the EEC's Council of Ministers in 1980 using indirectly the British Aerosol Manufacturers Association (BAMA) to reduce chloroflurocarbon gases in aerosol manufacturers' products. The British government's Department of the Environment requested the BAMA to get its members' compliance to reduce these gases by 30 per cent in less than two years. The Association put pressure on members and non-members alike to use alternative propellants and to measure the factory inputs of the offensive gases as ways to ensure compliance (Grant, 1983: 37–9, 63).

Of special importance to the British state in recent years is the promotion of manufactured exports. Public agencies that involve export-oriented economic management have returned to a relationship with industry associations that the British state severed at the conclusion of the Second World War. In the 1970s the CIA represented chemical industry employers on the Economic Development Committee for the Chemical Industry and the Chemical Industry Training Board (Gill et al., 1978: 46). By the 1980s the CIA was involved in overseas trade missions with funding provided by the British Overseas Trade Board. A second association, BAMA, has entered into joint ventures with the Board for participation in foreign exhibitions with BAMA displaying manufactured products of its members. The British Plastics Federation has also done this, and in addition the Board provided resources for a comprehensive study of the industry in India (Grant, 1983: 21, 36, 50). Similarly in order to deal with the high-input energy prices to all British industry that (especially) put the British chemical industry at an international competitive disadvantage, the British state formed 'a tripartite National Economic Development Council energy task force in which a member of the CIA Council was the sole customer industry representative' (Grant, 1983: 23).

There is a less private interest government in the Netherlands, although in the area of quality control there was once considerably more. As in the case of Austria, Sweden and Britain there are instances of private interest government in price regulation. However, more recently the Dutch chemical industry has backed away from hard-nosed collective action, although it is not averse to milder forms of associative activities such as information collection and aggregation (Grant, no date, pp. 13–15).

In many ways private interest government in the German chemical industry is the most interesting of all. This very strong industry stands out even in the country that is the world's leader in the export of manufactured goods. By the end of October 1986 the Federal Republic of Germany replaced the United States as the globe's dominant exporter in total dollar amounts.

Germany, as a European state that exhibited delayed development in both economic and political spheres, sought to boost its position by collective action, and by the turn of the century hundreds of domestic industrial cartels were organized, including some in chemicals. Immediately after this process had matured the Germans took the lead in organizing international cartels. The earliest and strongest cartels were chemical ones. The Bismuth Syndicate, formed in the 1870s, is probably the best known of the early, well-organized groups. However, cartels were formed to cover trade in 'borax, quinine, acetic acid, potash, carbide and explosives'. This German industrial style of collective self-discipline for long-term domestic and global economic dominance persists successfully down to the present day (Mirow and Maurer, 1982: 14, 16–19, 117–21, 131–42, 144–5).

The trade association of the German chemical industry, the Verband der Chemischen Industrie (VCI) – the Chemical Industry Association, relies on an even more impressive and positive strategy to ensure its lead on the world economic state. The VCI is committed to a collective research effort as the mechanism to maintain its leading edge. To this end the Verband administers a multi-million-dollar research organization, the Fonds der Chemischen Industrie. The revenue is generated by a fee schedule based on members' sales (Grant et al., no date, p. 15).

As in Sweden the West German chemical industry is treated as an integral part of the policy-formulation process. The initial drafts of proposed legislation that are of relevance to the industry are reviewed routinely by the VCI. The Verband makes use, in this case, of the expertise located in its member firms. These experts, once mobilized for this purpose, become spokesmen for the industry at large (Grant et al., no date, pp. 18–19).

The public policy role of the VCI demonstrated itself clearly in one of the major public policy concerns of the first half-decade of the 1980s; namely, the health and environmental impact of new industrial chemicals. The federal government passed the Chemicals Act, Chemikaliengesetz, in 1980 and began enforcing it in 1982. Its major provision was a pre-*marketing* notification of new industrial chemicals. Although this is seen as 'the major governmental intervention into the German chemical sector' (Schneider, 1987: 21) the burden is not as heavy as it might have been. While this Act was being formulated in Germany a more stringent law was put together in the United States, namely the Toxic Substances Control Act (TSCA). The American legislation demanded an even greater intrusion in the US industrial chemical industry in that 'a pre*manufacture* notification program' (Biles, 1983: 39) is required.

The difference between the two countries on the same policy issue lies in the policy position of the most important BIA in each country. While the Chemical Manufacturers Association (CMA) was an outsider in the pluralist policy process that produced TSCA, the VCI was an integral part of German policy-making. The components of the Chemikaliengesetz were established before the legislature became involved. In this process a basic negotiation took place between the Verband and the state agencies. So important was the role of the BIA that an analysis of imputed policy reputation in its policy area listed it as almost as important as the enforcing ministry (Schneider, 1986: 407–11). A second example of close industry–state cooperation on environmental policy was the quick action of the German chemical industry to reduce voluntarily the use of 'fluorocarbons as propellants' (Stoel et al., 1980: 123).

It is in the area of the implementation of environmental protection of watercourses that the West German chemical industry has displayed weak and ineffective self-government, despite claims of heavy spending in this policy area. The demonstration of the ineffectiveness of the industry's efforts, ironically, was triggered by an event outside the country, but which almost immediately had a direct, dramatic impact on the West German ecology. On 1 November 1986 a major chemical spill from the Switzerland warehouse of Sandoz AG, one of the world largest chemical companies, emptied into the Rhine River. Hundreds of thousands of fish and eels died, and West German authorities were forced to shut down water works along the river, leaving many thousands of Germans without water for days.

What this incident did inside Germany was to focus attention on the safety procedures of German chemical plants along its watercourses. Whether because of bad luck or increased public awareness a number of German spills from the chemical industry were noticed quickly in the same month. Twelve days after the Sandoz spill Hoechst, one of the three largest industrial chemical companies in West Germany, leaked benzol chloride into the Main River. Meanwhile, on the very same day polyvinyl chloride escaped from a smaller chemical plant at Woldshut, Baden-Württemberg into the Rhine. On 21 November the second of the big three firms, BASF, discharged two tons of toxic chemicals into the Rhine from its huge plant in Ludwigshafen, Rhineland Palatinate. Finally, four days later the third of the big three industrial firms, Bayer, allowed a chemical disinfectant to escape from its waste treatment facility at Krefeld-Ürdingen, North Rhine-Westphalia.

By the first week of December the federal government and the public had lost all confidence in the joint industry–state commission on environmental accidents. The various chemical mishaps were viewed now as intentional events reflecting ordinary industrial practice, and the federal cabinet removed all six industry members from the commission. In their place the cabinet appointed technical experts who were to respond on the basis of their specialized expertise rather than as industry's self-selected regulators.

The failure of joint chemical industry/state regulation of environmental accidents is rooted in some factors that are found only in some countries such as West Germany. It is not unusual to find industries coping with collective regulation, whether public or private, by bargaining with regulators (Peacock, 1984: 3). In Germany environmental-protection regulation is or was a shared responsibility of the federal government, the Länder, and joint federal–industry bodies. This mixture of divided public authority and mixed public–private institutions encourages industry's reliance on negotiation, and avoidance of compliance while negotiations proceed. The industry's position is strengthened further by clauses that require the public authorities especially to show that their enforcement is both economically suitable and technologically possible. Given the complexity of administration, and the content of the regulations themselves, the burden of showing how environmental pollution can be prevented is placed on regulating bodies (Wagner, in Peacock, 1984: 118–20, 129–30, 132).

The current weakness of private interest government in domestic environmental protection policy in the West German chemical industry is reinforced by the weaknesses of the Chemical Workers Union, IG Chemie and its pre-1933 predecessor, the German Factory Workers Confederation, the Verband der Fabrikarbeiter Deutschlands – UFD. Both unions had and continue to have difficulties in organizing the big three chemical companies. By the late 1970s the IG Chemie organization represented only 40 per cent of the workers at Hoechst and Bayer, and around 65 per cent at BASF. This makes the union's demands for domestic environmental protection, especially worker safety, easier for the companies to ignore, especially since they know that their own workers would rather trade off health risks for current higher wages (Markovits, 1986: 267–326).

Canada in comparative perspective

Compared to the European countries Canada is less likely as a site for private interest government, but even here there are numerous instances of this phenomenon in the chemical industry. In addition, as we shall see below, the Canadian practice resembles far more the European model of private interest government based on informal rather than formal inter-party relations and marked by consensus rather than conflict.[5] Various chemical associations in subsectors such as plastics – especially processing, pesticides, fertilizers and basic chemical production – have even become involved in the shaping of industry structure by joining together in industrial research and development.

The Society of the Plastics Industry of Canada (SPI-Canada) set up the Canadian Plastics Institute (CPI), whose basic purpose is to defuse new technology among plastics processors in Canada. The idea for this institute grew out of a 1978 plastics subsector task force, which was one of twenty-three formed to develop solutions to long-term structural problems facing these subsectors. The major problem identified for the plastics industry was Canada's high trade deficit. Not only does the Institute provide one central location for technical reference for the plastics industry, it also attempts to influence research work done in universities and other institutes, so as to make it more responsive to the plastics industry, especially the plastics processors.

The Institute could not have been started or survived in its early years without extensive state financing. The CPI, which was born in May 1983 and officially began operations in January 1984, received $1 million in funding from both SPI-Canada and the federal government over the first five years of operations. In addition the government's Science Council of Canada awarded CPI a large contract to carry out a world-wide study of the emerging new technologies most applicable to the Canadian plastics processing and fabricating industry. The CPI has but one member, SPI-Canada, thus all of SPI-Canada's member firms are indirectly members. The Institute is linked to the state and SPI-Canada by more than finances. On its twelve-member Board of Directors is the manager of the Plastics, Rubber and Chemical Specialties Division, Resource Processing Industries Branch, Department of Regional Industrial Expansion (DRIE). The Secretary of the CPI is the president of SPI-Canada and the Canadian Standards Association is also represented on the Board.

A second case is the Canadian Agricultural Services Coordinating Committee, which has to advise it two expert committees composed of members who must be approved by the head of the Crop Protection Institute of Canada (CPIC).[6] These expert committees, one on pesticide use and the other on weed control, affect the agricultural chemical industry directly by establishing the direction for new and in-use chemicals on a regional basis. These guidelines affect their use in relation to amounts, frequency and place (Coleman and Jacek, 1981: 57–8).

Finally, in the area of research and development two research-oriented associations are committed to increasing the use of chemicals whose Canadian production exceeds consumer demand. Most important is the Sulphur Development Institute of Canada (SUDIC), which is a major example of such activity. The purpose of SUDIC is to develop and promote new uses of sulphur by increasing research and development into these new uses. SUDIC is supported by grants from the government of Canada and Alberta (50 per cent). Somewhat less important is the activity of the Potash and Phosphate Institute of Canada (PPIC), which gives financial support to research projects designed to promote or improve the use of potash and phosphates as fertilizers. This research absorbs 60 per cent of PPIC's annual expenditure. In the labour policy area SPI-Canada is active. The Human Resource Development Committee of SPI-Canada laid the groundwork for SPI-Canada to sign a joint agreement with Canada's Department of Manpower and Immigration, which eases the entry of qualified immigrants for the Canadian chemical industry and provides the support for establishing training programmes.

A critical policy concern in the general policy area of quality and health standards is the hygienic manufacture and safe handling of chemicals. In 1971 the Canadian Chemical Producers Association (CCPA) established the Transportation Emergency Assistance Plan (TEAP) which manages the response by which member firms assist police, fire and civil protection authorities in the event of transportation accidents involving chemicals. Later, in 1979, the CCPA participated in the establishment and continuing support through TEAP of the Transportation Emergency Centre of Transport Canada (CANUTEC), which provides a 24 hour a day emergency response and information service to combat problems relating to any dangerous product, not only chemicals. CANUTEC must often refer calls to TEAP. The CCPA, together with the Chlorine Institute (a North American organization) and the Crop Protection Institute of Canada (CPIC), reorganized TEAP in 1983 in order to assure a speedy on-the-scene response at any serious chemical transportation emergency in Canada (Johnson, 1985: 90–1; Hart, 1985: 142).

As a result of this 1983 reorganization of TEAP a grid of regional control centres was established across the country. For example, the CCPA established a TEAP regional response centre in southwestern Ontario which is part of a Distribution Emergency Response System operated by Dow Chemical Canada in Sarnia, Ontario (Johnson, 1985: 87–95). This system is supplemented by a North American grid called the Chlorine Emergency Response System (CHLOREP), which was set up by the Chlorine Institute based in New York City. The Institute's members produce gaseous and liquid chlorine or provide these producers with supplies. In the case of an accident involving chlorine the nearest chlorine producer responds to the problem. Returning to Dow Chemical Canada's Distribution Emergency Response System in Sarnia, in addition to it being the TEAP regional response centre in southwestern Ontario it also responds 'to bulk liquid chlorine emergencies under the Chlorine Institute's "CHLOREP" system in south-western and north-western Ontario' (Johnson, 1985: 95).

In the general policy area of macro-intersectoral policies Canadian chemical BIAs are important agents of the state in environmental policy, weights and measures, and energy conservation. Canada has its own special brand of tripartism in environmental policy involving the national government, the ten provincial governments and chemical BIAs. Labour, environmental and consumer associations are excluded from the usual process. The Canadian Chemical Producers Association and the Canadian Manufacturers of Chemical Specialties Association are integral to the process. The tripartite task forces have drafted guidelines dealing with such chemicals as PCSs, Mirex, PBBs and chlorofluorocarbons (Ilgen, 1985: 581–6).

Also critical to Canadian environmental protection is pollution that emanates from the United States somewhere along their 3000-

mile border. Dealing with such trans-border problems is the responsibility of the International Joint Commission (IJC). In 1972 the Great Lakes Water Quality Agreement was signed between the United States and Canada. This Agreement is implemented by the IJC through two international boards, the Great Lakes Water Quality Board and the Science Advisory Board. The latter Board has representatives of state agencies, the public and business including a technical director of the Canadian Producers Association. This Board develops recommendations on research matters related to the identification, evaluation and resolution of current and predicted problems concerning water quality in the Great Lakes.

The two other federal government objectives pursued and largely met through the use of chemical business interest associations are the conversion of Canadian weights and measures from imperial to metric, and energy conservation as a result of the 1974 oil shock. Most BIAs have metric and energy conservation committees set up to mobilize their firms in reaching these policy objectives. In most sectors, with the exception of spotty resistance by a few retailers, the conversion from the traditional weights and measures has been smooth and relatively rapid. Similarly, the Canadian Industry Programme for Energy Conservation – led by sectoral task forces chaired by BIA staff in general manufacturing, farm and industrial equipment, machinery, non-prescription medicines, plastics and wood products – set a 1980 goal of 12 per cent for the increase in industrial energy efficiency. This target was actually exceeded by 3.4 per cent.

The most important use of an industrial chemical BIA as an agent of public policy is in the area of petrochemical plant capacity and Canadian industrial strategy. This is a complex area involving the Canadian Chemical Producers Association (CCPA), which has problems coping with one related issue, free trade with the United States, but which successfully helped the federal government solve a difficult regional rivalry in the large and growing petrochemical sector. From the national capital the second problem is always the more important one, since the federal government's success is measured internally by the maintenance of economic peace among the regions of this very large and diverse country.

In order to deal with the relatively small economies of industrial scale in Canada the Conservative government, elected on 4 September 1984, sought to expand substantially its markets by establishing a free trade zone in North America. In this strategy it was fully supported by the Royal Commission on the Economic Union and Development Prospects for Canada. This Commission was appointed by the Liberal government of Pierre Trudeau in November 1982, and headed by his former Finance Minister, Donald S. Macdonald. In addition, Republican President Ronald Reagan in the United States was happy with a proposal that fitted in completely with his free trade, market-led view of international economic affairs.

This proposal has met with strong support from the key over-arching BIAs in Canada. This included the two representatives of Canadian large business, the Business Council on National Issues (BCNI) and the Canadian Chamber of Commerce (CCC), as well as the two voices of Canadian small business, the Canadian Federation of Independent Business (CFIB) and the militant organization that split off from it, the Canadian Organization of Small Business (COSB). In addition, the multi-sectoral voice of Canadian industry, the Canadian Manufacturers Association (CMA) reversed its century-old position on requiring trade barriers between the two North American countries. Finally, the two major business associations representing those engaged in international trade, the Canadian Export Association (CEA) and the Canadian Importers Association (CIA), also supported trade liberalization. A number of sectoral and subsectoral BIAs supported the free trade initiative as well. Thus the government had and still has majority support among organized business. In the minority area there were a number of sectoral and subsectoral associations that stood to lose out in open borders, including three that cover part of the industrial chemical sector. These include the Society of the Plastics Industry of Canada (SPI-Canada), which is more responsive to its small and medium-sized Canadian-owned processor members rather than its largely foreign-owned members who produce bulk resins, the Rubber Association of Canada (RAC) whose small plants are expected to face stiff American competition despite the early Canadian technological lead in producing synthetic rubber and the Canadian Textiles Institute (CTI), producers of man-made fibres. Finally, and of greatest importance to us, are those associations that were split on this issue because their individual product domains were expected to be affected very differently under a free trade regime. Here we find the Canadian Chemical Producers Association (CCPA).

The CCPA organizes three distinct subsectors, each with varying economic characteristics that would affect the Canadian subsectoral position under free trade rules. The largest and fastest-growing subsector is petrochemicals, which now represents almost two-thirds of the Canadian industry and is nearly three times larger than it was a decade ago. There were fifty-two petrochemical plants operating at capacity level in the mid-1980s. This subsector strongly supported more secure and open access to the US market, where tariffs on Canadian petrochemicals now average between 10 and 20 per cent, in order to keep up its utilization of capacity so that economies of scale can be realized continually. The second subsector, inorganic chemicals, was a mixed bag of fifty-eight plants with a positive trade balance of about C$33 million in 1985, less than one-sixth of the surplus produced by the petrochemicals. About 25 per cent of the Canadian industrial chemical industry is this type of production. The CCPA was not able to come up with a consistent position on bilateral free trade for inorganic chemicals because some plants are world-scale works with state-of-the-art technology and a large production capacity for serving both Canadian and foreign markets, while others are of a small producing scale at relatively high unit cost for Canada's small market. The third subsector, the organic and specialty chemicals, is generally seen as a weak, timid one. It runs at a low utilization of its capacity, about two-thirds, and has a growing trade deficit of C$1½ billion. On the other hand, by 1983 the CCPA was looking forward to reduced tariffs in the 1980s as a way to facilitate entry of Canadian organic and specialty chemicals into foreign markets.

While the CCPA was having trouble coming to grips with Canada's free trade proposal to the United States, the Association was solving a contentious problem among the petrochemical producers themselves. The dramatic increase in petrochemical manufacturing capacity produced and planned by the end of the 1970s pitted the three key provinces of Alberta, Ontario and Quebec against each other as a result of the recession of the early 1980s. Especially bitter was the rivalry between the natural gas-based producers in Alberta, Canada's centre of hydrocarbon production, and the oil-based manufacturers in Quebec. The former were convinced that market forces would allow them to close down their rivals. However, given a large block of Quebec seats in the House of Commons, over three and a half times as many parliamentary seats as in Alberta, and the Quebec voters' propensity to pay close attention to short-term economic conditions, it was politically unthinkable that the Quebec-based Progressive Conservative federal government would close down any Quebec plants no matter how uncompetitive they were. Given the latter government's close ties with Quebec, the national political leaders were not well-placed to deal with the rivalries among Canadian firms. The federal government had its hands full just trying to keep a fragile peace among provincial political leaders.

At this point the CCPA became a forum for a resolution of this issue. The groundwork was laid, however, by a growing consensus about the future of the Canadian petrochemical industry. As early as 1981 a bipartite business-labour task force set up by the federal and provincial ministers of industry saw petrochemicals as the major export opportunity for the 1980s, an opportunity that needed to be followed up given the deterioration of Canada's manufactured goods balance of trade which reached a 'net deficit of $12 billion' by 1978, 'four times what it had been only eight years previously' (Consultative Task Force on Industrial and Regional Benefits From Major Canadian Projects, *Major Canadian Projects Major Canadian Opportunities: A Report*, 1981: 43). By 1984 government, industry and labour began to believe that all existing and planned petrochemical facilities had a future in a world where by some time in the 1990s Canada would be the largest exporter of petrochemicals in the world. This was to be accomplished by carving up the US and Japanese markets among the Canadian regions plus state action to ensure success.

The Western Canadian plants, especially those in Alberta, are sitting on top of huge supplies of very low-cost natural gas. Thus they have a clear world feedstock advantage coupled with the fact that the plants are new. Their market strategy, apart from supplying Western Canada with their products, is to expand their markets in the Western United States and Japan. Indeed joint Canadian/Japanese ventures have already taken place. The major drawback to these plants is transportation costs, which the federal government has responded to by deregulation of transport as a way to reduce these costs to the industry.

The Ontario plants, mainly located in southwestern Ontario, are large, world-class, state-of-the-art ones within very short distances, less than 500 miles in all cases, of their Ontario and Northern Midwest American customers. Thus in this US market they have a clear advantage over their Texas Gulf Coast competitors. The Ontario plants' main disadvantage was their high dependence on high-priced oil feedstocks. However, the federal government, through its programme of financially supporting conversion to natural gas feedstocks and then by contributing to the cost of moving this feedstock to Ontario, has reversed this disadvantage.

The policies to protect the viability of the Quebec producers are the most complex of all. First of all their markets are Quebec manufacturers, the four Atlantic provinces and adjacent Northeastern United States, especially New England. Traditionally this part of the American hydrocarbon market suffers from the highest commodity costs in the US, with the least secure supply, visibly demonstrated during the two 1970s oil shocks. Thus, the oil-based Quebec plants have less competition in their American market compared to the other two Canadian petrochemical centres. The federal Liberal government in the 1970s provided security of supply by extending the oil pipeline from Western Canada to Quebec. In addition one firm, Petromont, is jointly owned by Union Carbide and the provincial government, and another Union Carbide plant that produces ethylene has been assured that the federal government will cover any near-term losses in addition to the $15 million it has already paid out.

The American model

Pluralist analysts believe that the state-centred 'European model' of active state intervention in the economy and corresponding strong institutionalized interest associations is inimical to industrial competitiveness, high economic growth, low levels of unemployment and stable, minimal rates of price inflation. Rather the 'American model' with its market-oriented weak public authority approach, is believed to be much more suited to reach the goals stated above. The 'American model' is supposed to result in weak, poorly organized associations of interests dominated by the ideas of freedom, voluntary action and spontaneity, and strictly separated from public authority.

Our analysis of American industrial chemical associations will show that such an image is an overly gross distortion of a very complicated yet understandable American reality. In the United States, state agencies are much more powerful than is commonly believed outside that country. Their rules and regulations place often relatively severe limits on industry's freedom of action.[7] Business interest associations often spend an inordinate amount of time fighting debilitating battles against state rules and regulations in the law courts. If we seek an answer to America's prosperity we are probably better off placing emphasis on its traditionally large population subject to internal free trade, rich resources capable of exploitation over the past 100 years, a temperate climate, a highly educated labour force and possibly technological spinoffs from the state-financed defence and space programmes.

Business interest association legal challenges to administrative regulations are commonplace in the United States. Indeed many US associations, such as the Chemical Manufacturers Association (CMA), make it clear that law suits against state agencies are a normal part of their activity.[8] In the case of industrial chemicals many legal challenges to the Occupational Safety and Health Administration (OSHA) have come from BIAs over recent years.

Sometimes these suits are successful even for poorly staffed BIAs. In 1973 in the case of Dry Color Manufacturers Association v. Brennan, 486, F. 2d 98 (1973), this association with virtually no permanent staff argued that the OSHA had erred in promulgating emergency regulations on an 'Emergency Temporary Standard on Certain Carcinogens'. The court agreed that 'the statement of reasons preceding the regulation was inadequate' and the Standard was voided. A year later in another case where the OSHA had issued a permanent standard for the carcinogen chemical MOCA, the Polyurethane Manufacturers Association successfully argued in court that 'the regulation was illegal because the proposed final regulation had been issued without awaiting completion of a report of an advisory committee made up of business, labour and environmental medicine specialists set up to recommend on the final regulation'. The court demanded that the OSHA allow more comments to be made for a period before the regulation was to go into effect (Kelman, 1981: 55 and 62).

The most impressive instance of private interest government in the United States is the complex arrangements established by the industrial chemicals subsector to deal with the hazards of chemical manufacture, storage, transportation and disposal. This is done through the Chemical Manufacturers Association (CMA), which purports to represent both American and Canadian companies producing basic industrial chemicals. Traditionally, the CMA counts as its members the large firms in the subsector and thus has averaged at least 90 per cent of the American and Canadian subsector production among its members. At first blush the CMA looks to be a traditionally pluralist association with its strong political emphasis on lobbying the American Congress through its Government Relations Committee, its Government Relations Department and Vice-president, and most especially its Congressional visit programme with its designation of Congressional Liaison Representatives. But it is the array of seemingly technical programmes dealing with the control of the hazards of the modern industrial chemical industry that provide a rapidly growing instance of private interest government in the bosom of pluralist America. These hazards have produced huge political and legal risks for even this powerful industry, risks that increasingly move business firms into a strong non-market form of political–economic governance under the guise of arrangements that are labelled as politically unimportant or merely technical.

The Chemical Manufacturers Association (CMA) has set up the National Chemical Response and Information Center (NCRIC) to provide the public and emergency response organizations with information about chemicals, and advice or assistance during chemical emergencies. Although the Center is clothed with the pluralist language of voluntarism, from the beginning the threat of government intervention has pushed the CMA, together with the Chlorine Institute, the Fertilizer Institute and the National Agricultural Chemical Association, into becoming agents of public policy. Later legislation at both the national and state levels has increased the public character of these business associations and given them increasingly, in an indirect fashion, the characteristics of private interest government.

The oldest component of NCRIC is the Chemical Transportation Emergency Center (CHEMTREC), which was started by the CMA in 1971 under pressure by the National Department of Transportation (DOT). The Department in turn was pressurized by the public after a major train derailment that involved dangerous chemicals. Two choices to deal with transportation crises involving hazardous chemicals were possible: the national government could set up its own agency to deal with the problem or the CMA could deal with the problem. The government preferred that the industry handle its own problems, for two reasons. First, the CMA had in place a much better communications network with chemical producers. Second, and more importantly, the Department realized that the CMA would get much better cooperation from both the member chemical firms and the shippers of bulk chemicals. CHEMTREC worked so well that the Department of Transportation, in March 1980, designated it, in a formal agreement, as the central emergency response service for dealing with incidents involving the transportation of hazardous chemicals. CHEMTREC notifies DOT of major incidents through the US Coast Guard's National Response Center (NRC). In the DOT's own *Emergency Response Guidebook* it is CHEMTREC's telephone number that is listed. The NRC number is not listed in its own *Guidebook*. In addition, shipments of industrial chemicals in the United States 'must be accompanied by shipping papers that describe the hazardous material and contain a certification by the shipper that the material is offered for transport in accordance with the applicable "Department of Transportation" regulations'. On these shipping papers next to 'Emergency Phone Numbers' (in bold letters) are CHEMTREC (in caps) and its US and Canadian toll-free phone number (in bold letters), its local Washington, DC number and its number for callers outside North America. These numbers are manned by technical personnel seven days a week, night and day – a fact stated on the shipping papers. No government telephone numbers are found on the shipping papers (US Congress, 1986: 11, 156–7).

However, later national legislation turned a private interest agent of public policy into a strong private interest government. In 1985 the US government passed 'right to know' legislation. This legislation requires chemical producers not only to identify the contents of their products, but also to give a telephone number where they can be reached should an emergency arise related to their product. CHEMTREC's 24 hours a day, seven days a week centre in Washington fulfils the legislation. Since all of CMA's members are expected to be participants in CHEMTREC, their member firms need do nothing further. However, the chemical producers, mainly small ones with 10 per cent of the production, who did not belong to CMA before the 'right to know' legislation, found themselves in a difficult position. Given their small size it is uneconomical for them to maintain their own emergency centre. Thus, the imperative of the new legislation is forcing these remaining chemical producers to join the CMA and accept its standards.

Should the shipper and the producer not be able to respond quickly enough, normally defined as within two hours, to a transportation emergency, another NCRIC component comes into action. This is CHEMNET, which is a mutual-aid network between shippers, chemical producers and CMA-approved hazardous goods contractors that provides advice and hands-on assistance at the site of major chemical transportation crises. In order to meet the 'within two hours response time' in getting a chemical emergency response team on site, the CHEMNET system goes first to another CMA member, a non-producer of the shipped chemical in this case, who has the required personnel in the region of the chemical distribution accident. If all CMA companies are too far away then CHEMNET will put the shipper in contact with a nearby chemical contractor. The latter must meet the standards of the CMA and agree to maintain them by signing an agreement with the CMA. Only then does the chemical contractor receive the designation 'CHEMNET Contractor' and thus is allowed to participate in the system. Overall in this way no site in the United States is more than two hours removed from an emergency response team.

However, to make CHEMNET work, non-producers that respond to other companies' problems have insisted they be freed from any legal liability in acting as 'Good Samaritans'. In order to remove this legal jeopardy, special laws need to be passed by the American states. Of the fifty states, thirty-eight now have these 'Good Sam' laws, as they are called. Of major concern to the chemical industry is that two very large states, California and Ohio, have yet to pass this legislation. These laws are critical because they lay the foundation for the mutual-aid network of

CHEMNET. The major opponent to such laws is the Trial Lawyers Association. 'Good Sam' laws limit the grounds upon which these lawyers, representing clients who claim to have been adversely affected by chemical transportation accidents, can seek financial redress in the courts.

Supplementing and integrated into the CHEMNET system are three other networks. The Chlorine Institute, the BIA that organizes chlorine manufacturers, maintains CHLOREP, the Chlorine Emergency Response System to deal with chemical transportation incidents that involve chlorine dangers. This system uses NCRIC and its components to act as a central dispatcher in these problem situations. The second system is PSTN, the Pesticide Safety Team Network, set up by the National Agricultural Chemicals Association to deal with pesticide spills. Finally there is a set of 'external mutual-aid agreements' among 'the Channel industries, a concentration of chemical facilities along the Texas Channel . . . in Houston, Texas' (US Congress, 1986: 224).

A third key component of NCRIC are programmes that give training in emergency response. Programmes include emergency response team workshops that prepare chemical and carrier response team members to handle transportation incidents involving hazardous materials. Secondly, a lending library of audiovisual training programmes has been established, which includes training programmes that have been developed by the American government's Federal Emergency Management Agency, the Fertilizer Institute, which organizes the American fertilizer industry and the Chlorine Institute, among others. In addition, the CMA is developing a new videotape programme for chemical emergencies' first responders who are likely to be police, firefighters and emergency medical personnel. In addition, 'the Chlorine Institute provides training publications and films on chemical and physical properties of chlorine, appropriate regulations, and handling procedures for normal and emergency operations.' Other BIAs that offer 'training in hazardous materials and response procedures include the Association of American Railroads', which coordinates its own reporting system and databases on chemical rail accidents and chemical releases with NCRIC, the American Trucking Associations, the National Agricultural Chemical Association and the American Petroleum Institute which represents the American petrochemical industry. 'Some states such as Virginia base their "State Emergency Response Training Program" on industry resources and on CHEMTREC' (US Congress, 1986: 76, 140, 231, 233–4).

The final component of NCRIC is the Chemical Referral Center

(CRC), which is designed to enable all users of chemical products, whether transportation workers or the general public, to obtain safety and health information about chemicals. Calls from all over the United States can reach the Washington, DC centre toll-free, thirteen hours a day. This Center was set up in response to the 'right to know' legislation mentioned above. The CRC is able to tell the caller the name of the chemical company that produced the product, and the caller will be given the person and number to call in order to obtain the detailed information requested. Complementing NCRIC's activities is the new CMA programme called Community Awareness and Emergency Response (CAER). This programme was started under government pressure as a result of the chemical disaster in Bhopal, India in 1984. This pressure increased following an accidental gas release at Institute, West Virginia.

CAER involves a number of steps, the most crucial ones at the local level where there are chemical plants. Initially the CMA reviewed emergency planning efforts generally and then attempted to apply them to the chemical industry. CMA officials then wrote a work-book for chemical plant managers, who were directed to set up local coordinating groups. These groups are composed of these managers, local government officials, police, fire and rescue personnel, plus other interested community leaders and citizens. The job of these local coordinating groups is, first of all, to develop integrated emergency response plans composed of chemical plant emergency plans joined together with other existing local plans. In order to assist this planning process the CMA, besides developing a work-book, has also prepared a videotape, a 35-mm slide presentation and a brochure. In addition the CMA has been holding workshops and seminars for chemical plant managers and their associates.

Not only is this programme integrated into the plans of local government officials, but CAER is also being coordinated and implemented with American national agencies, especially the Federal Emergency Management Agency. This comprehensive programme to deal with chemical emergencies places an obligation on the chemical companies to inform the local community of details of chemicals manufactured or used at local chemical plants, something which has not been done before. Once again a major impetus to this programme, as with some others described above, is the 'right to know' laws, which produce an integrated industry response because of the fragmentation and inefficiencies that would result if individual companies each attempted to establish their own programmes. By February 1987 a total of 1500 American chemical plants were participating in the CAER programme and 800 chemical plant emergency plans were completed. In addition 200 plant managers had established local community coordinating groups which are integrating the local chemical plant emergency plans with community evacuation plans already in existence. Chemical plant managers, although reluctant to take on these additional duties, find themselves pressurized from two sources. First, the CMA which has announced publicly that 'material safety data sheets on commodities manufactured and stored in the community' are available to the public from plant managers, and secondly the state 'right to know' laws (US Congress, 1986: 60).

The final step of the chemical industrial process, that of waste disposal, is also dealt with in the form of private interest government by the CMA. In 1979 the CMA created a Hazardous Waste Response Center. This unit offers teams of industry experts to federal, state and local governments when advice is needed in dealing with waste disposal sites. Through a group of advisory response teams the Center has investigated a number of sites and submitted reports to the federal government's Environmental Protection Agency (EPA). The Center also prepared a manual on hazardous waste site management. There is good reason to believe that in this policy subarea, as in many environmental ones, collective industrial self-regulation, if it meets general governmental standards and goals, is often less costly, less time-consuming, and a more efficacious risk-management system than direct governmental regulation (Baram, 1982: 53–62).

The American industrial chemical industry is responding in a collective fashion to other public policy demands. Especially troublesome for the US chemical industry is the Toxic Substances Control Act (TSCA) of 1976. Its purpose is to ensure adequate testing of new chemicals, beyond the now existing 55,000 manufactured chemicals, in order to assess their possible adverse effects on human health and the environment, including economic ramifications. The EPA must be notified of these impacts before manufacture begins (Hadden et al., 1982: 10). The EPA is also concerned about the environmental impacts of the existing industrial chemicals, especially the large-volume ones, about fifty in number. These are produced by a number of firms. In order to avoid duplicate testing the CMA and the Chemical Industry Institute of Toxicology (CIIT) are testing, or plan to test, a large number of these as well as thirty-five other chemicals not in the top fifty (Blair and Bowman, 1983: 75). An advantage of using industry associations for this problem is that state agencies are more willing to accept test results because the actual manufacturing firms, with their obvious conflict of interests, are not involved in the research. In addition, the American Industrial Health Council (AIHC) is available to address these same concerns. Increasingly all these associations are responding not only to queries about existing commercial chemicals but also 'to some of TSCA's requirements' for new industrial chemicals (Harlow, 1983: 131).

While the American chemical industry has ventured to use business interest associations as agents of public policy in environmental policy areas it has relied on market mechanisms to deal with the industry-wide problem of global overcapacity, fierce foreign competition and depressed prices and profits. As we have noted, this is a problem that affects the petrochemical industry primarily; one that is especially important to the US since American production is about one-third of the world's total.[9] However, attempts to restructure the industry by American firms have failed to solve the US overcapacity problem. Weaker, inefficient firms have not left the market place because assessments of market forces are made by the state, which is not trusted, or by strong firms, which are believed to bias the policy analysis. In addition US laws on hazardous wastes and site clean-ups make it economically difficult to close down individual plants. In any event collective discussions on rationalization would put US firms at risk of violating US anti-trust laws, thus leaving the firms with no alternative except to try market-driven solutions (Bower, 1986: 15, 23, 46, 75, 181–2).

There is one way the American petrochemical industry has responded to world markets that has allowed the US petrochemical trade surplus to reach almost $8 billion in 1986, and shows realistic hope for the future. US firms are increasingly acknowledging the cheaper feedstocks, especially methane and ethane from natural gas, available in the energy-rich countries. These feedstocks produce primary products, that is aromatics and olefins, in energy-rich countries. These primary products are then transformed into the much higher value-added organic intermediates. Thus, the US tolerates trade deficits in benzene, an aromatic, and ethylene, an olefin. Over the long run the same will probably happen to trade in propylene, an olefin, that has a short-term trade surplus due to exports to Western Europe. In contrast, the US has major trade surpluses in three important organic intermediates, namely vinyl chloride monomer, styrene and toluene diisocyanate.

The ongoing good health of the American chemical industry is of critical concern to that country. It is an industry that employs over one million people, including approximately 150,000 professionals. In terms of trade balances it has the second best record of all US industries; this in a country with a huge trade imbalance in the 1980s. The chemical industry counts for almost 10 per cent of US manufacturing yet the industry worries about four major problems. First, it believes the American public is hostile to the industry. This attitude leads to a second problem from their viewpoint, a continued public policy thrust to protect workers, the public and the environment from chemical poisoning. Third, American petrochemical processors should expect greater and greater competitive pressure from the energy-rich countries with their inexpensive and abundant feedstocks and their new, efficient plants. Finally, the preceding two factors plus high US wage rates and the high value of the US dollar have led to a seemingly permanent tight profit squeeze (Committee to Survey Opportunities in the Chemical Sciences, 1985: 3, 209–10, 327–8).

It is within this context that we should note the efforts of American chemical BIAs to manage trade policy, especially trade with Canada which now runs in the latter's favour. As is the case with key overarching multisectoral BIAs in Canada, their counterparts in the US take the same position. At the same time, as in Canada, a number of sectoral and subsectoral BIAs support North American free trade. Thus, the American government, like its northern counterpart, has a strong majority support among organized business, whether counted by firms or associations. Again, there are a minority of sectoral and subsectoral BIAs that expect to lose out in an open border regime and thus opposed free trade, including some in the industrial chemicals sector. Overall, though, the US industrial chemical industry favoured free trade with more intensity than its Canadian counterparts.

Private interest government and organizational development

How is it that business interest associations can develop the properties of private interest government? The answer would seem to lie in the concept of the relative autonomy of the business interest organization. As the organization develops from a purely member-dependent voluntary association concerned with urgent, immediate problem-solving to one with a diversified and dependable financial, personnel and status resource base, and with an increasing attention to important, long-term planning, then the organization staff develops sufficient autonomy to influence both members and potential members. In this way the special interests of current association members are replaced by more general public interests.

There is another important characteristic, that is the difference between rules affecting only the members of the voluntary association and rules affecting also non-members but firms and individuals belonging to a certain category or domain; for example, firms in a sector or subsector. Thus any organization can regulate the behaviour of its members, whether it be a firm, a church, an aquarium club or a business association. But that is not what we would call private interest government. The latter concept refers to private associations creating regulations which have a similar status to state regulations, that is, regulations which are *generally binding* for a certain category in society (*allgemeinverbindlich*), for example, the whole industry, rather than just the members of the association. The means by which such regulations become generally binding is of secondary importance.

In examining the actual relationship of resources to the development of private interest government outputs, we first must identify the components of this type of development. Whatever indicators we use they must involve the wielding of authority. There would seem to be four levels of private interest government outputs. The first involves coercive relations with members. The BIA may place demands on its members by asking them to perform certain actions such as signing public statements, refusing to collaborate with certain state agencies or engaging in lockouts. The organization may also reprimand, fine, suspend or expel members for acting contrary to association policy. Second, the BIA may be represented on a para-state body that wields public authority in such areas as social security, regional development, industrial policy, standardization of products, prices and wages. Third, the association itself may administer, directly or through its subsidiaries, public policy programmes in policy areas such as regional development, industrial policy, vocational training, quality control and price controls. Finally, a BIA may be involved in producing 'monopoly' or 'authoritative' goods such as commercial arbitration; enforcement of quality standards; certification and licensing; registration of patents; negotiation of binding agreements on minimum wages and conditions; negotiation of binding agreements with customer and supplier organizations on prices, supplies and conditions; enforcement of health and safety contracts; and control of members on such things as competitive practices and hiring-and-firing practices. We are concerned especially with the intensity of this monopoly activity and the proportion of organizational effort spent on the provision of monopoly goods.

The measurement of diversified and dependable resources involves at least four properties of BIA resource dependence: formalization of membership obligations in dues payments, stability of BIA

Table 7.1 *Nature of relationships between private interest government outputs and types of resources in the industrial chemical processing industry*

	Types of resources										
Outputs	Resource dependence on members	Increase in office staff	Resource dependence overall	Stability of BIA income	Formal consultation in legislation	Intensity of state access	Resource dependence on state	Legal to be consulted by state	Members taken to court	Expulsion of members	Special levy in last five years?
Intensity of institutionalization	Indirect −0.138 (P)	Indirect −0.299 (NP)	Direct 0.110 (NP)	Indirect −0.038 (NP)	Direct 0.202 (P)	Direct 0.373 (P)	Direct 0.180 (P)	Direct 0.102 (P)	Direct 0.259 (P)	Direct 0.658 (P)	Indirect −0.400 (NP)
Intensity of public policy administration	Indirect −0.465 (P)	Direct −0.027 (P)	Indirect −0.097 (P)	Direct 0.161 (P)	Direct 0.115 (P)	Direct 0.224 (P)	Direct 0.210 (P)	Direct 0.420 (P)	Direct 1.000 (P)	Direct 0.409 (P)	Indirect −0.458 (NP)
Provision of monopoly goods	Indirect −0.238 (P)	Indirect −0.215 (NP)	Indirect −0.097 (P)	Direct 1.000 (P)	Indirect −0.127 (NP)	Indirect −0.051 (NP)	Direct 0.368 (P)	Indirect −0.469 (NP)	Direct 1.000 (P)	Direct 0.560 (P)	Direct 0.398 (P)
Proportion of organizational effort spent on the provision of monopoly goods	Indirect −0.354 (P)	Indirect −0.127 (NP)	Direct 0.093 (NP)	Direct 0.42 (P)	Indirect −0.249 (NP)	Indirect −0.043 (NP)	Direct 0.433 (P)	Indirect 1.000 (NP)	Indirect 0.9289 (P)	Direct 0.042 (P)	Direct 0.065 (P)
Intensity of member sanctioning	Indirect −0.778 (P)	No relationship 0.000 (—)	Indirect −0.243 (P)	Direct 1.000 (P)	Indirect −0.492 (NP)	Indirect −0.702 (NP)	Direct 0.590 (P)	Indirect −1.000 (NP)	Direct 1.000 (P)	Direct 0.833 (P)	Direct 0.706 (P)
Demands placed on members by the BIA	Indirect −0.254 (P)	Indirect −0.374 (NP)	Indirect −0.169 (P)	Direct 1.000 (P)	Direct 0.167 (P)	Direct 0.191 (P)	Direct 0.172 (P)	Indirect −0.405 (NP)	Indirect −1.000 (NP)	Direct 0.261 (P)	Indirect −1.000 (NP)

P = predicted relationship; NP = opposite of predicted relationship, gamma coefficient.

income, trends in organizational bureaucratization and state recognition. High dependency on members should reduce the likelihood of private interest government while payments and staff given by the state to the BIA should increase the incidence of this form of private government. Overall resource dependence should not be related to this phenomenon. As the BIA is able to extract forcefully financial resources from its members, it should increase its private interest government outputs. It is able to do this by imposing special levies on its members, by taking members to court if they do not pay their dues and special levies, and expelling them for the same reason. Third, the stability of BIA finances should be related to the type of outputs discussed above. Finally, the intensity to which the association is given state access and recognition – that is, political institutionalization – should lead to more private government activity. Especially important here should be formal consultation in drafting legislation and representation before parliament and parliamentary committees.

In general three resource measurements – the lack of resource dependence on members, resources provided by the state and the ability of BIAs to expel members – are related consistently to the provision of private interest government outputs (see Table 7.1). It would seem that the ability of the organization to obtain grants from state agencies is critical to the organization's ability to free itself from dependence on the usual membership fees. In addition, the actual occurrence of the expulsion of member firms for not paying their dues or periodic special levies reinforces the financial resource independence of the organization and its consequent ability to take on governmental properties and problem-solving capacity.

Looking at the relationship from the output side it is clear that the intensity of public policy administration is best related to the properties of diversification and dependability of BIA resources. In only one instance out of eleven does the predicted relationship not hold. This level of private interest government, the actual private administration of state policies, is the level that is most clearly related to the very resources that should be expected to nurture private interest government in the industrial chemicals industry.

Conclusion

There are a number of important positive findings that arise out of this chapter. Private interest government has developed in different countries in response to important public policy problems. The two most general problems facing the industrial chemical industry are the problems of industrial overcapacity, especially in petrochemicals, and the negative response of both public authorities and public opinion to the perceived dangers to human safety and environmental degradation from the industrial chemicals industry.

Business interest associations have the potential to deal with long-term overcapacity problems, but in both Europe and the United States the rules of the political game make such efforts difficult and probably illegal. In both cases the rules of the game are rooted in nineteenth-century liberal conceptions of small independent firms competing on the basis of price. These rules become manifest in 'the Treaty of Rome's Articles 85 and 86' for Europe and the Sherman Antitrust Act for the United States (Bower, 1986: 57–61, 85). In contrast the Canadian government allowed, and indeed facilitated, industry cooperation. Of course the task before the Canadian industry was easier. The problem was and is orderly expansion, not contraction. The future shines brightly with large markets in the northern United States and Japan, while their competitors in those areas face much higher feedstock prices with dwindling petrol reserves beginning in 1987 in one case and a lack of domestic oil and gas in the other.

The second major problem facing the industrial chemical industry is the health and environmental issues. Increasingly major concerns about the impact of industrial chemicals on the health of workers, consumers and local communities near chemical plants and major transportation corridors are being raised by the public, unions and state officials. On the other hand, a number of chronic major environmental problems of a global nature, such as acid rain and smog which comes from the energy consumption of the general public (Scarrow, 1972: 261), or the greenhouse effect which results from energy production cannot be laid at the door of the industrial chemical industry. In contrast at least a couple of major problems can be laid there, namely pesticide contamination of natural populations and the deterioration of the ozone layer.

Countries do vary in major ways in their attempts to deal with environmental chemical hazards through collective self-regulation. Although Europe is viewed by most people as having a tradition of state interference with industry compared to the United States, the relationship between industry associations and the state is more relaxed, informal and cooperative than in America (Brickman et al., 1985: 310; Vogel, 1986: 146; Kelman, 1981). Canada represents an intermediate case more like Europe than the United States (Ilgen, 1984: 3 and 1985: 587) yet the pressures for policy harmonization in North America abetted by the enthusiastic endorsement of continental free trade by Prime Minister Mulroney and President Reagan have pushed Canadian policy-making closer and closer to the American model.

Is such a trend good for Canada? Should not Europeans emulate the American political economy model of the 1980s? There are good reasons not to. The current allure of the American model is based on the seeming strong recovery from 1970s stagflation and the 1982 recession. Yet the apparent American economic progress hides a harsh reality. Unemployment has declined markedly over the past five years but the quality of jobs, both to the workers and to the economy as a whole, has also declined. The loss of well-paid positions for skilled workers in industrial production is more than being offset in aggregate numbers by poorly paid unskilled and semi-skilled service jobs (Dumas, 1986: 18–23). The causes of the American industrial decline beneath the bravado of Ronald Reagan's economic liberalism and the shallowness of open unemployment rates are beyond the scope of this chapter. All that need be said here is that the adoption of American practices solely because of its apparent economic success in the 1980s would be a mistake.

But what about the social regulation of the industrial chemical industry? Are not the American people better off than others because of the tough hostile behaviour of American bureaucrats towards industrial pollution? There is no doubt that American laws towards the industrial chemical industry are relatively harsh, and public officials are demanding. Comparatively the US industry has to spend greater sums of money in order to comply with governmental regulations. Costs are also increased by combating new governmental initiatives legally and politically at every turn. Because of the uncertainty of long political and legal battles coupled with inter-agency and inter-governmental conflict with American federalism, the financial burden of the social regulation to US chemical firms is the highest in the world. Yet the products restricted and the actual protection given to the population is no more than that given in Europe or Canada. Thus, once more the positive surface impression gives way to a more thorough and pessimistic assessment of American reality (Brickman et al., 1985: 310–14).

What has been missed in the policy analysis of business–government relations in the US is the vast amount of confusion that exists within both governmental organizations and business firms, and the great financial expense this entails. Inter-agency conflict within the federal government alone produces enormous delays and expensive legal battles. Meanwhile innovative research, development and production within business firms is 'put on hold'.

Within firms themselves conflicts develop between government liaison employees and line managers and shopfloor workers. Often, because of the complex horizontal and vertical differentiation within business firms, managers and government relations specialists cannot control subordinates on pollution control matters because the former do not know what is really occurring at the production level (DiMento, 1986: 114–57). In addition, as government environmental policy changes – as it frequently does in the US because of new legislation and displaced presidential administrations – corporations that manufacture industrial chemicals frequently adjust their corporate organization to deal with these changes (Hadden et al., 1982: 85–96).

Certainly a case for policy harmonization can be made. The solution would seem to lie not in mechanically choosing one country as a policy model to be superimposed on the rest of the world. Rather it seems more reasonable to work out global problems of the chemical industry on the global level through international BIAs on their own, or as insider participants in international organizations organizing member states. For illustration we can refer to two important problems: the unintended and harmful effects of pesticide residues and the impact of chlorofluorocarbons (CFCs) and other chlorine-containing chemicals on the earth's ozone shield.

Unintended pesticide residues are an unfortunate side-effect of our modern agricultural production. This problem travels across national borders first because residues remain in the food consumed in a country outside of the nation of origin and also because of the migration of species in the food chain across national boundaries. In order to deal with the first problem the Food and Agriculture Organization (FAO) and the World Health Organization (WHO) have set up the *Codex Alimentarius* Commission. Its key committee is the pesticides residues committee and its working subcommittees or groups.

It is at these levels we see involvement by the chemical industry through the Groupement International des Associations Nationales de Fabricants de Produits Agrochimiques (GIFAP) and its twenty-one national associations representing the pesticide industry. GIFAP 'has observer status at meetings of the Codex pesticide residues committees'. It usually has the largest technical staff present at these meetings, sometimes in excess of thirty people. Statements made by GIFAP 'are then published in the official report of the session'. In addition the BIA takes an active part in the discussion within the working subcommittees (Boardman, 1986: 45–69). A second global environmental problem involving chemicals is the production of CFCs and other chlorine-containing chemicals which are responsible for a massive depletion of the stratospheric ozone layer which protects our skin, eyes and immune system from the bombardment of extraterrestrial ultraviolet radiation (Morone and Woodhouse, 1986: 76–85). Clearly all or most major industrial nations must be committed to the reduction and eventual elimination of these manufactured chemicals. Early on some national BIAs played an active constructive role. The public authorities in Austria have always been among the most pro-control in the world, but the voluntary Austrian Aerosol Association (Österreichische Aerosol Vereinigung – ÖAV) has played an important role in this policy development and implementation. A somewhat contrasting case is Britain, which on the state level in international discussions has been balking at a freeze and then phase-out of industrial CFCs. However, internally the British Aerosol Manufacturers Association (BAMA) has been used by both the British government and the EEC to reduce CFC use in aerosols.

The contrast between BIA reluctance to become involved in industrial restructuring because of a probable violation of anti-collusion rules of the game, and the numerous instances of BIA action to meet environmental challenges confirms Graham Wilson's argument that 'different policy areas will involve business in different types of politics' (1985: 8). The politics of pollution control is fundamentally different from the politics of economic restructuring. Cooperation is viewed as desirable in one case but not in the other.

Cooperation, not competition, is desirable in an industry based on high-risk technology. BIAs can provide training, help and information to industrial operators so that as potential environmental accidents start to take shape the front-line personnel can act to separate manually parts of the industrial technology so that unfortunate interactions do not take place. Unfortunately, unexpected and threatening interactions are common in the chemical industry which, when they produce a series of failures, leads to disaster (Perrow, 1984: 101–22). BIA efforts at helping chemical operators to cope with the unexpected can reduce accidents and the severity of those that occur, but some incidence of accidents is inevitable.

In these situations, as in other policy areas, the ability of BIAs to solve problems in an industry such as chemicals depends on the way interests are organized, the type of resources they have and the institutional constraints they must live with, at least for the short and medium term. Policy solutions in the modern economy need this type of explanation more than they do 'state-centric theories' or 'national interest' calculations (Hall, 1986: 230). State institutions and the legal rules of the game are important, but their interaction with organized interests appear to offer the best understanding of outcomes in the political economy.

At the same time we reject those analyses that posit solutions to industry-wide problems by looking to creative firm management alone (Bower, 1986: 207–24) or new and rediscovered forms of work organization such as 'flexible specialization' (Piore and Sabel, 1984: 258–63). So much industrial analysis has been done in recent years to show that the individual firm has limited room to manoeuvre in dealing with overcapacity of one or more products in an integrated, complex chemical plant or firm. In addition, exit is made more difficult by environmental protection rules that demand that closed-down chemical sites, which often include adjacent hazardous waste dumps, be cleaned up and made environmentally safe for future use.

To solve collective problems of the industry BIAs must develop a special relationship between resources and outputs. For a BIA to develop the properties of a private interest government it must diversify its financial resources. It will remain undeveloped if its sole source of finance is regular membership dues. Most important is the ability of the organization to extract financial resources from the state. By so doing the organization can make itself more insulated from its members and thus develop greater relative autonomy.

For associations 'to engage in long-term strategic thinking about the problems facing their industries' (Coleman and Grant, 1984: 209), these resources must be predictable and be from varied sources. The association's staff must be free from quickly responding to the member's immediate concerns. Dealing with the urgent problems of member firms must not be allowed to displace the effort to cope with enduring important problems of the industry. Instead, the staff must be able to insulate itself from immediate member demands by having dependable, alternative sources of finance and symbolic state support. Indeed, by having such resources the association staff should develop the ability to control member behaviour sufficiently so as to act as implementing agents of public policy.[10]

Notes

1. The concept of 'private government' has at least two basic meanings. The second broader understanding will not be used in this chapter. This alternative usage sees hierarchy and authority in private organizations as 'government' and

distinct from market relations which are identified as the normal hallmark of the private sector. The objectives and uses of this authority are secondary to its exercise. Thus Gilb (1981: 464) defines a private government as 'the government of a business or industrial corporation, a bank or other financial institution, a trade association, church, or any other limited-purpose organization'.

2. In developing these five major areas of policy I have built this classification upon my previous analysis (Jacek, 1987).

3. Here I am referring to some de facto imputation of public status by state authorities to an organization that could exist as a private element of civil society. For an analysis in keeping with my meaning see Offe, 1981: 123–58.

4. The public functioning of BIAs in 'crucially important industrial sectors' outside of the chemical industry is perhaps more well known. Energy and coal-mining in West Germany is one example. The state's attempt at coal import substitution for the electricity industry was frustrated until 'coal mining and the electrical supply industry associations' took responsibility to make the state policy work (Schmitter and Streeck, 1981: 103). This has led to continual major reductions in the importation of foreign coal. By the first half of 1986 almost 90 per cent came from inside Germany. This is especially remarkable since by the last quarter of 1986 foreign coal, mainly from Poland, was only half as expensive as domestic West German coal.

5. This tendency is the same as noted by Ilgen (1984, 1985) in his study of the regulation of toxic substances in Canada.

6. Until 1986 the CPIC was known as the Canadian Agricultural Chemicals Association. However, the Association felt that the media and Canadian public opinion were concerned excessively with the environmental impact of agricultural chemicals. This, together with depressed prices for Canadian crops, led to a significantly lower demand for these chemicals, a 15 per cent drop in 1986. In order to boost sales by presenting a more positive image for the pesticide industry the Association changed its name to the CPIC. In contrast, environmental concern is currently at a lower level in the United States. Thus, the National Agricultural Chemicals Association there has not changed its name.

7. I am certainly not the first to recognize the generally unappreciated hostility of American state agencies to business compared to the working partnership between business and the state in Europe. James Q. Wilson in his 'Forword' (Kelman, 1981: ix–x) also notes this 'counterintuitive' difference. Also see Vogel (1986: 267) and Brickman et al. (1985: 247).

8. In the 'How We Work' section of a CMA publication this association states that it presents 'sound positions to governmental bodies'. However, 'If this approach fails, CMA is prepared to go to court' (Chemical Manufacturers Association, 1981: 11).

9. US petrochemical exports are also very important. The major destinations of these exports are Western Europe (30 per cent), Latin America (21 per cent), non-Japan Asia (12 per cent), Japan (11 per cent) and Canada (12 per cent). Major imports now come from Western Europe, Canada and Japan (Bower, 1986: 93).

10. Mayntz (1983: 140) calls them 'agents of policy implementation' and points out the use of 'these indirect modes of political control' may have less to do with national differences of culture or policy styles than with the specific policy problem and the ability of a specific government to deal with it directly.

References

Baram, Michael S. (1982) *Alternatives to Regulation: Managing Risks to Health, Safety and the Environment*. Lexington Books.

Bengtsson, Lors, Eriksson, Ann-Charlotte and Sederblad, Per (1984) 'The associative action of Swedish business interests: the Swedish employers' confederation and centralized collective bargaining in 1980, 1981 and 1983', IIM/LMP 84-24, Wissenschaftszentrum, Berlin.

Berger, Suzanne and Piore, Michael J. (1980) *Dualism and Discontinuity in Industrial Societies*. Cambridge University Press.

Biles, Blake A. (1983) 'Harmonizing the regulation of new chemicals in the United States and in the European Economic Community', in George W. Ingle (ed.), *TSCA's Impact on Society and Chemical Industry*. American Chemical Society. Chapter 4, pp. 39–65.

Blair, Etcyl H. and Bowman, Carlos M. (1983) 'Control of existing chemicals', in George W. Ingle (ed.), *TSCA's Impact on Society and Chemical Industry*. American Chemical Society. Chapter 6, pp. 67–78.

Boardman, Robert (1986) *Pesticides in World Agriculture: The Politics of International Regulation*. Macmillan.

Bower, Joseph L. (1986) *When Markets Quake*. Harvard Business School Press.

Brickman, Ronald, Jasanoff, Sheila and Ilgen, Thomas (1985) *Controlling Chemicals: The Politics of Regulation in Europe and the United States*. Cornell University Press.

Chemical Manufacturers Association (1981) *CMA: What It Is, What It Does*.

Coleman, William D. and Grant, Wyn P. (1984) 'Business associations and public policy: a comparison of organisational development in Britain and Canada', *Journal of Public Policy*, 4 (2): 208–35.

Coleman, William D. and Jacek, Henry J. (1981) 'The political organization of the chemical industry in Canada', a research paper presented to the annual meeting of the Canadian Political Science Association, Halifax, Nova Scotia, 27–29 May.

Committee to Survey Opportunities in the Chemical Sciences (1985) *Opportunities in Chemistry*. National Academic Press.

Consultative Task Force on Industrial and Regional Benefits from Major Canadian Projects (1981) *Major Canadian Projects, Major Canadian Opportunities*. Government of Canada.

DiMento, Joseph F. (1986) *Environmental Law and American Business: Dilemmas of Compliance*. Plenum Press.

Dumas, Lloyd J. (1986) *The Overburdened Economy*. University of California Press.

Dyson, Kenneth and Wilks, Stephen (eds) (1983) *Industrial Crisis: A Comparative Study of the State and Industry*. St Martin's Press.

Gilb, Corinne, L. (1981) 'Public or private governments?', in Paul C. Nystrom and William H. Starbuck (eds), *Handbook of Organisational Design*, Volume 2: *Remodelling Organisations and their Environments*. Oxford University Press. Chapter 21, pp. 464–91.

Gill, Colin, G., Morris, Richard and Eaton, Jack (1978) 'Employer organisation in the UK chemical industry', *Industrial Relations*, 9 (1): 37–47.

Grant, Wyn P. (1983) 'The organization of business interests in the UK Chemical Industry', IIM/LMP 83-3, Wissenschaftszentrum, Berlin.

Grant, Wyn P. (1984) 'Is corporatism necessarily interventionist? A discussion of the Thatcher government in Britain'. Paper presented to the Conference on 'The

Ständestaat in Historical and International Perspectives', Ludwig Boltzmann Institut für Historische Sozialwissenschaft, Salzburg, Austria, 12–13 April.

Grant, Wyn P. (no date) 'The overcapacity problem in the European petrochemicals industry' (unpublished manuscript).

Grant, Wyn P., Paterson, William and Whitston, Colin (no date) 'Government–industry relations in the chemical industry: an Anglo-German comparison' (unpublished manuscript).

Hadden, Susan G., Howe, Richard S. and Rohlich, Gerard A. (1982) *Environmental Regulation and the Chemical Industry*. University of Texas.

Hall, Peter A. (1986) *Governing the Economy: The Politics of State Intervention in Britain and France*. Polity Press.

Harlow, Dan R. (1983) 'Impact on corporate structure and procedure', in George W. Ingle (ed.), *TSCA's Impact on Society and Chemical Industry*. American Chemical Society. Chapter 9, pp. 121–32.

Hart, Christine E. (1985) 'The transportation of hazardous products', in J.H. Shortreed (ed.), *Dangerous Goods Movements*. University of Waterloo Press. pp. 129–44.

Ilgen, Thomas L. (1984) 'Between Europe and America: regulating toxic substances in Canada', in F. Thompson (ed.), *Regulatory Regimes in Conflict: Problems of Regulation in a Continental Perspective*. University Press of America. pp. 1–30.

Ilgen, Thomas L. (1985) 'Between Europe and America, Ottawa and the Provinces: regulating toxic substances in Canada', *Canadian Public Policy/Analyse de Politiques*, 11 (3): 578–90.

Jacek, Henry J. (1987) 'Business interest associations as private interest governments', in W.P. Grant (ed.), *Business Interests, Organizational Development and Private Interest Government: A Study of the Food Processing Industry*. de Gruyter. pp. 34–62.

Johnson, R.W. (1985) 'Canadian chemical industry actions and concerns related to the reductions of risk in the transportation of hazardous materials', in J.H. Shortreed (ed.), *Dangerous Goods Movements*. University of Waterloo Press. pp. 87–97.

Katzenstein, Peter J. (1984) *Corporatism and Change: Austria, Switzerland and the Politics of Industry*. Cornell University Press.

Katzenstein, Peter J. (1985) *Small States in World Markets: Industrial Policy in Europe*. Cornell University Press.

Kelman, Steven (1981) *Regulating America, Regulating Sweden*. MIT Press.

Marin, Bernd (1983) 'Organizing interests by interest organizations: associational prerequisites of cooperation in Austria'. *International Political Science Review*, 4 (2): 197–216.

Marin, Bernd (1985) 'Austria – the paradigm of liberal corporatism?', in W.P. Grant (ed.), *The Political Economy of Corporatism*. Macmillan. Chapter 4, pp. 89–125.

Markovits, Andrei S. (1986) *The Politics of the West German Trade Unions: Strategies of Class and Interest Representation in Growth and Crisis*. Cambridge University Press.

Mayntz, Renate (1983) 'The conditions of effective public policy: a new challenge for policy analysis', *Policy and Politics*, 11 (2): 123–43.

Mirow, Kurt Rudolf and Maurer, Harry (1982) *Webs of Power: International Cartels and the World Economy*. Houghton Mifflin.

Morone, Joseph, G. and Woodhouse, Edward J. (1986) *Averting Catastrophe: Strategies for Regulating Risky Technologies*. University of California Press.

Offe, Claus (1981) 'The attribution of public status to interest groups: observations on the West German case', in S. Berger (ed.), *Organizing Interests in Western Europe*. Cambridge University Press. Chapter 5, pp. 123–58.

Peacock, Alan (ed.) (1984) *The Regulation Game: How British and West German Companies Bargain with Government*. Basil Blackwell.

Perrow, Charles (1984) *Normal Accidents: Living with High-Risk Technologies*. Basic Books.

Pestoff, Victor (1983) *The Organization of Business Interests in the Swedish Chemical and Drug Industry*. University of Stockholm.

Piore, Michael, J. and Sabel, Charles F. (1984) *The Second Industrial Divide*. Basic Books.

Scarrow, Howard A. (1972) 'The impact of British domestic air pollution legislation', *British Journal of Political Science*, 2 (3): 261–82.

Schmitter, Philippe C. and Streeck, Wolfgang (1981) 'The organization of business interests: a research design to study the associative action of business in the advanced industrial societies of Western Europe' (revised and extended version). Berlin: International Institute of Management.

Schneider, Volker (1986) 'Tauschnetzwerke in der Politikentwicklung: Chemikalienkontrolle in der OECD, EG und der Bundesrepublik Deutschland', *Journal für Sozialforschung*, 26 (4): 383–416.

Schneider, Volker (1987) 'The structure of policy networks: a comparison of the "chemical control" and "telecommunications" policy domains in the Federal Republic of Germany'. Paper presented to the ECPR joint workshops session, 10–15 April, Amsterdam.

Stoel, Thomas B., Miller, Alan S. and Milroy, Breck (1980) *Fluorocarbon Regulation: An International Comparison*. Lexington Books.

Streeck, Wolfgang and Schmitter, Philippe C. (1985) 'Community, market, state – and associations? The prospective contribution of interest governance to social order', in W. Streeck and P.C. Schmitter (eds), *Private Interest Government: Beyond Market and State*. Sage. Chapter 1, pp. 1–29.

Traxler, Franz (1983) 'Structure and functions of the business interest associations in the Austrian chemical sector'. Paper presented to the 'Organization of Business Interests' sector conference on the chemical industry, Rome, 21–22 April.

US Congress, Office of Technology Assessment (1986) *Transportation of Hazardous Materials*, OTA-SET-304. Washington, DC: US Government Printing Office, July.

Vogel, David (1986) *National Styles of Regulation: Environmental Policy in Great Britain and the United States*. Cornell University Press.

Wagner, Elaine (1984) 'The environment of negotiation in the Federal Republic of Germany' and 'The bargaining process in the Federal Republic of Germany', in A. Peacock (ed.), *The Regulation Game: How British and West German Companies Bargain with Government*. Basil Blackwell. Chapter 7, pp. 116–30 and Chapter 8, pp. 131–54.

Wilson, Graham K. (1985) *Business and Politics: A Comparative Introduction*. Macmillan.

8

Chemical Unions in the United Kingdom and Germany

Colin Whitston

Business associations and firms are not the only organized interests within the chemical sector, and account must also be taken of the workers within the industry, and the ways in which they define and pursue their independent goals. Here it is somewhat more difficult to characterize sectoral features since the legal and institutional structures of industrial relations are chiefly shaped by national rather than sectoral concerns. Nevertheless, the internationalization of the industry, its characteristic production methods, and high degree of capital intensity make comparative study a fruitful possibility. This chapter concentrates on the organization of worker interests in the UK and the Federal Republic of Germany, their responses to changing production and market conditions, and how they 'fit' into areas of public policy, nationally and internationally.

The very different historical experiences of British and German labour mean that, although facing many similar problems, they do so within very different constraints. In general a contrast can be drawn between the 'voluntarist' nature of British industrial relations, its multi-unionism, and its emphasis on workplace organization, and the legally regulated system in Germany, the predominance of industrial unionism, and the institutions of codetermination. These differences are important, and are reflected in both the structure and conduct of unions but, in both countries, specific sectoral features are also important in understanding patterns of influence and power in the industry.

Although the chemical industry developed far earlier in the UK than in Germany, effective trade union organization was a relative late-comer in both countries. As a result both sets of organizations faced a concentrated and bureaucratically organized industry. For the UK this meant that craft tradition was less important in the formation of industrial relations in the industry than in manufacturing in general. Both the state and employers have had an influence on bargaining arrangements that has fostered a relatively cooperative ethos in the industry. The German industry appears at first sight to conform closely to the national model, but there are ways in which the institutions common to the country at large have been modified in chemicals, to the disadvantage of the unions. These features are traced through the organization and conduct of collective bargaining, and their connections with some important areas of public policy, and in particular with the regulation of health and safety issues.

International organization

To understand fully the role played in the industry by organized labour, collective bargaining must be put in the wider context of the structure of markets and the influence of government policies. By their nature unions take as their starting point for organization and policy-making the interests of their national membership. Likewise their political aims, and avenues of political influence, are concentrated at the domestic level. In an industry which has always been highly international this limitation has always been recognized, and chemical unions have been prominent in international labour organizations. Traditional union internationalism, however, has tended to concentrate on questions of 'unfair' competition through the use of cheap labour, and an emphasis on supporting union organization in developing countries. The spread of multinational corporations, which plan production on a regional or even global scale, presents new problems. These are compounded in the case of the chemical industry both by an emerging regional organization of production within Europe on the part of many major companies, and by the growing importance of the European Community itself as a political centre of economic and social decision-making. Union responses to these new challenges have been relatively muted to date, and this has been due to both organizational and political problems.

The development of international trade union organization in the postwar years has been inextricably bound up with the cold war, starting with the break-up of the World Federation of Trade Unions (WFTU) initiated by the British and American federations after WFTU attacks on the Marshall aid programme. The International Confederation of Free Trade Unions (ICFTU) established in 1949 was strongly anti-communist, and anti-communist sentiment remained an important feature of the organization even after the exit of the American unions in 1969 (Coldrick and Jones, 1979: 14). Additionally pressures in respect of national interests intervene in ways which make the ICFTU a highly political body. For instance, the international department of the TUC (which according to some accounts provides about one-third of the ICFTU's income) 'is not particularly accountable and some claim that the links between . . . [it] and the Foreign Office are stronger than those with the international department of the Labour Party' (Eaton, 1978: 10). Similar political pressures have shaped the international organization representing chemical workers. The present hegemony of the International Confederation of Chemicals, Energy and General Trades Unions (ICEF) was not established until 1976 when its rival, the International Federation of Petroleum and Chemical Workers, was exposed as a 'front organization' for the CIA (Coldrick and Jones, 1979: 82).

The ICEF has established regional organizations, and its European Coordinating Committee operates industrial committees which include oil, rubber and chemicals. IG Chemie, the German chemicals union, plays a leading role in funding and operating this committee, and the coordinating committee for the EC is chaired by the head of its International Relations Section. Hostility to communist unions remains strong in the ICEF, particularly amongst the Germans, while at the same time not all British unionists involved are convinced that it is not under some influence from the Americans. These complications have little impact on day-to-day relations between the constituent unions, but they have served to complicate the development of a consistent European policy, and representation within the EC.

Formally speaking, contact with the European Commission is coordinated by the European Trades Union Confederation (ETUC). This body was established in 1973 out of a sometimes uneasy alliance between national federations comprising ICFTU affiliates, Christian and communist unions. Essentially the ETUC acts as a coordinating and advisory body with no direct bargaining functions. It is dedicated to creating a 'social space' for unions within the EC through the promotion of laws and directives in the labour interest, and by advocating Keynesian-style growth policies.

The adoption of the sixth directive on hazardous substances was of particular concern to chemicals unions because of its implications for occupational health monitoring, and is a good example of the issues which the ETUC can pursue at European level. However, this case also illustrates some of the difficulties of European coordination, especially within chemicals.

As with national federations the ability of the ETUC to develop expert advice depends on the input of unions organizing in a specific field. To this end it has established a number of industry committees, but in chemicals political differences have prevented such an initiative. Individual members of the ICFTU are organized into trade secretariats, such as the ICEF, and these secretariats fear both a diminution of their influence within Europe, and the development of a European 'parochialism'. In addition ETUC policy is to open membership of its committees to unaligned or WFTU unions equally with ICFTU affiliates. In chemicals IG Chemie has been consistently hostile to the establishment of any committee which would include the Italian communist-led union, the CGIL. As a result there is no unified committee covering chemical workers.

Although the ETUC monopolizes direct representation to the Commission, during consultations over biotechnology officials at DGXI held discussions with members of the ICEF's European chemicals committee. Although there is no direct evidence that this disunity affected the final shape of the directive, the potential weaknesses in the unions' ability to exercise pressure are clear. Some British union officials expressed regret over these divisions, but maintained that overlapping membership of many bodies, and the relatively small numbers of people involved, lessened the possibility of serious mistakes being made. Nevertheless, there is a certain irony in the fact that the ICEF occupy the same building in Brussels as the European Trade Union Institute (ETUI), which is the main research arm of the ETUC, but that the two organizations have no communications between them.

Political rivalry is not, however, the only factor retarding the development of trade union activity at the European level. In dealing with the international activities of the major corporations formal contact with the EC is often the least important priority, since the immediate priorities are most often determined by national considerations and the immediate issues of collective bargaining. Where issues intersect with public policy the national legislature is not only the immediate focus of political activity, but also where unions have the best established influence. This is most obviously the case on questions of labour law or economic and wages policies, but it also applies to those areas where European initiatives are important.

The case of toxic substances is again a good illustration of the process. While the TUC secured advice from interested unions during the development of the sixth directive, and pushed their views through the Social Committee and the European Parliament, in many ways the crucial stage lay in the exact form the enabling legislation would take in the UK. At this stage the unions had access to the formal structures of consultation established by legislation, as well as contacts within parliament.

As questions of trade policy, overcapacity and investment policies assumed increasing importance in the wake of the recession of the early 1980s other problems emerged for union coordination which related neither to political affiliation nor organizational divisions, but to differing assessment of national interest. An important case was the overcapacity problem in petrochemicals and the challenge of the new Middle East producers. The French and Belgian industries played a leading part in attempting to form an EC-sponsored 'crisis cartel', and the British unions, extremely worried by job losses and import penetration, backed the idea. While being far from fervent protectionists the unions blamed government for gross mismanagement of North Sea feedstocks, and were deeply suspicious of the commitment of the leading multinationals to UK-based production. IG Chemie took a much more relaxed view, both on trade and external investment by its own industry. By their own account they remained unenthusiastic about the proposed cartel, despite formal adherence to an international union declaration in its favour and, amongst British unions at least, were seen as operating delaying tactics on behalf of the 'free trade lobby' in discussions over the new producers.

The differences between the British and German unions are rooted in their very different domestic circumstances. In Britain the unions had been deeply committed to a planned use of North Sea feedstocks to develop a leading role for UK petrochemicals, and had put a lot of political effort into a tripartite strategy for the industry. In Germany IG Chemie had long held a view of economic priorities which fitted far more closely with the industry's aggressive export orientation. To understand these differences, and their implications for international cooperation, it is important to look at the history of union organization and interest formation in the two countries.

The organization and representation of union interests in the UK chemical industry

Union organization

The development of trades unionism in chemicals has been a protracted and uneven process, and it was not until after the war that it saw significant growth in coverage and density, especially in the white-collar sector. That the industry rarely makes headlines in the labour pages of the press reflects a history of relatively cooperative relations with employers, but this does not mean an absence of conflict, or an unambiguous commitment of employers to collective bargaining. The unions made important gains in wages and conditions during the long period of above-average growth in the industry during the 1960s and 1970s, but the recession of the early 1980s opened a more difficult period in relations with the employers. Particular problems have emerged in respect of work organization following restructuring in the industry, and a new willingness on the part of employers to minimize or evade union organization in some sectors.

A survey in 1967 found that the British chemical industry had 'a much lower proportion of workers with some skill . . . than any industry other than food and tobacco', and the 1981 census showed under 10 per cent of the industry's workforce was craft trained while over 30 per cent were unskilled process workers (Hunter et al., 1970: 209; Elias, 1985). In the circumstances it is hardly surprising that industrial relations on the manual side are dominated by the two major general unions, the Transport and General Workers Union (T&GWU), and the General, Municipal, Boilermakers and Allied Trades Union (GMB). Craft workers, predominantly engaged in plant maintenance and erection, are distributed among a number of unions, but with the highest concentration in the Amalgamated Engineering Union (AEU), and the electricians' union, the EEPTU. Despite their relatively small numbers craft workers have played an important role, with a history of militancy in some regions, and are currently central to employers' strategies towards skill flexibility.

At the end of the Second World War union membership in the industry was under 30 per cent for all employees, while less than half of ICI's workers were organized, but the next thirty years saw strong growth, reaching over 58 per cent for the industry as a whole in 1979 (Gill et al., 1978: 87; Bain and Price, 1983: 14). This growth was most marked among manual workers (83 per cent), but also made deep inroads into technical and administrative workers. Since there is no industrial union for chemicals, estimates of membership size and distribution are difficult. However, ICI conceded a closed shop for its manual workers in 1968, and Brown estimated that 55 per cent of workers were covered by such agreements (Brown, 1981: 53). As union membership and density are known to increase with establishment size it is clear that unions are firmly established in the leading sectors of the industry.

White-collar employment has been the fastest growth sector of the industry, rising to about 41 per cent by 1981. The range of jobs under this category is large, ranging from technicians in production, through clerical and managerial posts, and this produced a spread of union organization, in many cases originating in staff associations or other forms of 'house unions'. Despite the large numbers involved, therefore, membership density of independent unions remained low until the late 1960s. A survey in 1964 found a membership density of only 3.7 per cent compared to 12.1 per cent for manufacturing as a whole (Hunter et al., 1970: 207; Bain, 1967: 27). At the time of the Donovan Commission no independent union had gained recognition from any major chemical employer with the exception of limited representational rights in certain plants (Gill et al., 1978: 156).

The Donovan Commission highlighted employer hostility to union recognition as a major cause of weak union organization amongst non-manual workers in manufacturing, and certainly this trait was shared by chemical employers. As late as 1973 ICI stated that it 'did not welcome unions for white collar workers because they fundamentally did not share the company's interests and objectives' (Gill et al., 1978: 181).

Nevertheless, by 1981, 41 per cent of white-collar workers in this industry were organized (Brown, 1981: 52). The reasons for this turnaround are varied. Inflation and the impact of wages policies and productivity bargaining on differentials seemed to have played a part (Gill et al., 1978: 164), but government measures to facilitate recognition were also important. The Chemical Industries Association (CIA) made several attempts to organize a national framework, but disagreements amongst employers led to a compromise agreement in 1973 which led to wider recognition, but on a company basis rather than an industry basis (1978: 166). While the organizations involved are a mixture of clerical unions and white-collar sections of the manual unions the Association of Scientific, Technical and Managerial Staff (ASTMS) was the largest single union for white-collar workers in the industry until its merger with TASS (formally the Technical and Salaried Staff section of the engineering union) to form Manufacturing, Science and Finance (MSF). Its membership in chemicals is now part of one of the largest, and potentially most influential, unions in the country, capable of organizing across current boundaries of skill and expertise. The union is referred to here by its pre-amalgamation title, ASTMS.

Despite this rapid growth white-collar membership remains far less evenly distributed between firms than manual unionization. For example, Esso still does not recognize a staff union, and in BP membership is only 28 per cent. By contrast the site employing the majority of another major chemical company's white-collar workers has a membership level of 76 per cent, rising to 88 per cent in non-managerial grades (Income Data Services (IDS) Report 445, March 1985: 5).

The major unions in chemicals nationally organize some four million workers right across industry, commerce and the public services, and are a major political force in both the TUC and the Labour Party. Their political orientation is varied, and has changed continually over the years. Currently ASTMS, the GMB and the T&GWU are counted on the centre-left of the TUC. The engineers and electricians were for a long time in open conflict with the General Council over their approach to industrial relations legislation, and are closely associated with the right wing of the union movement in their aggressive pursuit of the 'new realism' in collective bargaining, which led to the expulsion of the EEPTU from the TUC in 1988.

These political differences are important, but they should not be exaggerated in terms of the conduct of collective bargaining, or of the general economic role of trades unionism. The expulsion of the electricians from the TUC has made no appreciable difference to the conduct of industrial relations in the industry. There would be little dissent amongst them from an opinion expressed by Lord Cooper of the GMWU in 1962 that 'We consider that industrial relations would be significantly improved if more firms regarded Trade Unions and collective bargaining as valuable instruments in promoting the objectives of the firm to everybody's benefit' (Eaton and Gill, 1978: 313). Nevertheless the three core unions, ASTMS, GMB and T&GWU, have developed the closest sectoral links, having been drawn together in tripartite work on the industry's Economic Development Committee, and in 1978 they formed the Chemical Unions Council, a loose association of the principal officers concerned, to act as a focus for information exchange and policy coordination at the sectoral level.

Employers' organization and collective bargaining

The Chemical Industries Association differs from that in Germany as the functions of trade and employers' association are united in a single body. However, not all members of the CIA are party to central industry-wide collective bargaining, and this arrangement of 'conforming' and 'non-conforming' membership dates back to the virtual collapse of the Chemical Employers' Federation in the inter-war years, largely due to the independent line taken by ICI, by far the largest employer.

Given this peculiarity the usual incentives for membership of an employers' organization must be reconsidered in the case of chemicals. Gill, Morris and Eaton, in the only comprehensive study of industrial relations in the industry published in Britain, argue the importance of 'Critical . . . state intervention . . . in the formation and expansion of employers' organization in chemicals', and the 'desire of employers to collectively influence or participate in state industrial relations policy' (Gill et al., 1978: 32). The role of state intervention goes back to the First World War, and the recommendations of the Whitley Council in 1917 which encouraged national bargaining. Equally important was the role of labour controls and arrangements with unions to reduce conflict during both wars, which were particularly pronounced in chemicals.

Gill et al. add to this the impact of increasing government intervention in the economy and labour law since 1945, and particularly the promotion of tripartism in a range of bodies from the National Economic Development Council to the Health and Safety Executive. Thus despite the weak (and declining) importance of national wages structures employers retained a strong incentive for collective organization (1978: 48). This is an important point, but such influences are by no means limited to chemicals. Also, as Gill et al. readily admit, the normal economic incentives may include, as Flanders argued, a strong political element aimed at securing 'Union assistance in managerial control . . . for the sake of gaining employees' consent' (1978: 32) which has proved extremely important in the negotiation of change at crucial periods for the industry. The commitment of the state to orderly collective bargaining and the promotion of interest representation at national level is also, of course, by no means a constant, and the government is currently engaged on the opposite tack, disorganizing and undermining union legitimacy. Nevertheless, what has already been said on white-collar recognition, and the pervasive influence of the state on the structure and conduct of industrial relations, is as essential a background for employers as for trades unionists.

There is no unified system of collective bargaining for workers in the chemical industry, but there are three sets of national Joint Industrial Councils which are of importance despite their limited coverage. The largest of these is the Chemicals and Allied Industries Joint Industrial Council (CAIJIC) for process and ancillary workers, covering 50 per cent of the industry's workforce. It is composed of three CIA nominees, and the Unions' Signatory Committee, representing the T&GWU, GMB and USDAW, and sets minimum wage rates, holidays, premium rates, among other things. Since the 1960s these agreements have been supplemented by bargaining at plant or company level, which has led to substantial wage drift with industry rates accounting for between 62 and 72 per cent of real earnings over the past ten years. The Drug and Fine Chemicals Joint Conference operates on similar principles, and the CIA–Craft Union Agreement negotiates craft differentials on the basis of the CAIJIC settlements. Non-conforming members of the CIA negotiate independently with a mixture of company-and plant-level agreements. The largest company involved is ICI, which employed between 25 and 30 per cent of the industry's manual workers throughout the 1970s. Taking these different practices together plant or company bargaining is the most significant determinant of earnings for up to 74 per cent of workers (Brown, 1981: 8). There are also wide variations in shift working and the use of payment-by-results systems, as well as in 'fringe benefits' such as service holidays and sick pay arrangements. Nevertheless, national negotiations remain important, with the CAIJIC agreement being regarded as a useful 'marker' in terms of wage competition (Grant, 1983: 15), as well as in the resolution of broader questions, such as productivity dealing in the 1960s, and negotiations over the working week and operating flexibility currently.

Equally important, the structure of collective bargaining presents both difficulties and opportunities for the unions in developing an approach to the sector as a whole. Fragmentation can obviously inhibit the development of coherent policy, and aggravate differences arising from competitive forces. However, the central role of the CIA in the industry, as well as the involvement of the same sets of union officers in all important bargaining units, means that in practice there is more coherence in union–employer relations than simple structures would suggest, although the network remains somewhat looser for craft workers, and even more so for the white-collar section.

Both the cohesion and the flexibility of this system of bargaining underlay the relative stability of industrial relations in the industry during the growth years of the 1960s and 1970s, and is well illustrated by the development of productivity bargaining. This form of bargaining became crucially important in the late 1960s under the twin spurs of government modernization policy, reflected in the work of the National Economic Development Office (NEDO), and pay restraint. Problems of international competitiveness had long been a concern of many managements, and the 'Fawley Agreement' reached by Esso in 1960 was in many ways a pacemaker for manufacturing industry as a whole. For the chemical industry in general, however, and particularly for conforming companies, the origins of productivity bargaining are to be found in a union–management visit to the USA under the auspices of the Chemicals Economic Development Committee to investigate manpower utilization, and the effects of economies of scale.

A report of the visit, issued in 1967, found that while differences in plant size accounted for most of the productivity gap between the two countries, 'Changes in manning practices could still increase production by a quarter' (*The Times*, 7 April 1967). The main conclusions emphasized the need for greater flexibility in the use of labour, especially across craft boundaries, the rigidities of some incentive schemes, as well as the desirability of a 'career' structure to integrate blue- and white-collar conditions. Many of these issues remain unsolved, and are further discussed below, but resultant changes in collective bargaining are still important in understanding the nature of industrial relations in the industry.

Productivity bargaining in the 1960s and 1970s typically involved a downgrading of industry-wide collective bargaining. In chemicals the process did involve a greater role for in-plant union organization, but it also produced new forms of centralized control, giving new meanings to relations between union officers and the CIA. In 1967 the CIA signed a 'framework agreement' with twelve unions which, for the first time, brought together process and craft workers in plant-level negotiations over productivity. A Joint Standing Committee on Productivity was also established, to advise companies, and with the responsibility of monitoring and authorizing deals reached by conforming members (IDS Report 37, 1968). This body was on the whole effective in limiting wage drift, and two reviews, in 1971 and 1972, checked bargaining fragmentation and reinforced the role of union full-time officers (IDS Reports 115, 1971; 137, 1972).

An early and influential agreement was introduced by BP at its Grangemouth plant in 1968, abolishing the 'clock', putting maintenance on a shift system, virtually eliminating systematic overtime, and introducing a salary structure for weekly paid staff. Management estimated that the agreement led to a reduction of man-hours per unit of output of 15 per cent without any reduction in absolute labour costs, representing 'A fantastic change in labour organization' (*The Times*, 26 July 1968). The usefulness of the Standing Committee was also emphasized when a non-conforming member, Associate Octel, signed a very similar agreement in 1972, including company-wide monitoring of plant agreements (IDS Report 137, 1972). The major non-conforming member also introduced a Weekly Staff Agreement (WSA) in 1968, buying out some restrictive practices, introducing a measure of craft flexibility under union control, weekly staff status, and a redundancy scheme.

The longer-term results of these schemes are difficult to assess. For example, while the WSA in ICI led to an increase in output per head rising from 7 per cent a year to 17 per cent between 1970 and 1972, account must also be taken of technical changes, and a reduction in the company's workforce between 1972 and 1973 from 102,500 to 89,000 (Gill et al., 1978: 95, 96). These problems were common to most of the schemes, and by 1972 the gains in productivity were running up against market restraints, and the impact of inflation. Nevertheless, one leading official maintains that this period raised chemical workers from the mediocre to the relatively well paid. An important sectoral peculiarity was also apparent, compared to manufacturing as a whole, in the conduct of productivity bargaining. The national negotiator for the GMB at the time of the framework agreement stressed the important conjunction of an 'Employers' Association with high standing . . . firms which have funds . . . and unions . . . accustomed to talking over a wider field than pay and conditions' (Wigham, 1968).

It would be easy, but wrong, to see in these relationships an uncomplicated and cooperative pattern of industrial relations. While relationships are more developed than in many sectors of British industry, companies are insistent on keeping union influence within strictly defined limits. The development of productivity bargaining was acceptable not only as a process of negotiated change, but also a way of fending off demands for extended and formalized involvement of employees in major areas of decision-making. While accepting the 'Importance being attached to the philosophy of participation', and the idea that productivity deals 'Provide an excellent opportunity for this development' (IDS Report 137) the employers in the industry have long been fierce opponents of any proposals for the extension of industrial democracy. In the period following the recession of the early 1980s, therefore, the process of change, while still relatively smooth, has not been so marked by cooperation.

Bargaining and recession

The impact of recession after 1980 had an immediate impact on the pattern of bargaining in the industry, with the settlements between 1981 and 1983 well below the level of inflation. This coincided with sharp cuts in employment of about 20 per cent, with BP for example shedding 40 per cent of its labour force, while at the same time productivity rose by 11 per cent. Profits recovered well in 1983, and this generated some new investment, but with continuing overcapacity in many sectors this has been concentrated on improving the competitiveness of existing plant rather than on expansion. While the recovery in chemicals was strong compared to manufacturing as a whole, another slump in 1985 hit the industry particularly hard. Output fell over the first two quarters of 1986, and while production recovered to the level of the first quarter of

1985 by the end of the year, growth rates remained sluggish. The impact on employment levels in this second downturn were negligible compared to the early 1980s, but basic rates were increasing faster than average earnings until the upturn in late 1986, indicating a strong influence by the trade cycle (IDS Report 498, 1987).

The longer-term impact of recession has been felt in new pressures on work practices, flexibility and staffing levels, particularly within the petrochemicals sector where overcapacity has been greatest. In this area the cooperative methods of negotiated change adopted in the 1960s have been noticeably absent, and many companies have attempted to move rapidly, spurred on both by competition and the reduced bargaining power of the unions (IDS Study 322, 1984). This has been particularly striking in some areas where changes have been introduced without union agreement, or have involved a weakening of union organization.

The construction of the Shell/Esso complex at Mossmoran was particularly important in this respect. The development was planned to take advantage of North Sea ethane supplies in the production of the basic 'building block' of the plastics industry, ethylene. Long before its completion, however, a chronic crisis of overcapacity developed in the sector on a world basis, and the new plant became a centre of management determination to restructure working practices, and their 'revolutionary' proposals soon created widespread anxiety amongst the unions. Neither company recognizes unions on the site, and labour has been recruited on the basis of maximum flexibility. Arrangements at this 'greenfield site' have been used to put pressure on workforces at other sites.

Shell used the new capacity to engineer a fundamental restructuring of operations at its Carrington plant outside Manchester. In April 1985 management announced the future closure of the ethylene cracker and the subsequent general reorganization of production. As a result an agreement was negotiated designed to save 500 of the 1200 jobs at the site. As part of the arrangements a twelve-member joint 'redeployment unit' was established with the aim of ensuring 'A future for Carrington, to ensure there are no compulsory redundancies, and to involve management in these objectives' (*Chemistry & Industry*, 5 August 1985). The plan included radical changes in grading structures and work patterns, replacing three agreements covering process, craft and staff employees with a single 'salaried technician' scale. The company did not attempt to combine job losses with pay cuts, and the new scale ran from £6741 per year at the bottom to £14,247 at the top.

With consolidated shift and disturbance allowances the lowest grades in plant processes earned £12,234 and the highest £16,563 (IDS Report 455, 1985).

However, while the unions were thus closely involved in the implementation of the plan, and indeed in negotiation over it, the shape of the 'survival package', and its presentation to the workforce and their families, was firmly under management control. This was clear since the deal was made 'on the premise that the "new Carrington" is effectively a greenfield site with complete flexibility of labour'. The company had considered insisting on a single-union agreement, but eventually settled with the four main unions on site – the engineers, the electricians, the T&GWU and ASTMS. Four unions lost recognition, and the ASTMS membership was removed from the company's national negotiating procedure (IDS Report 455, 1985).

Changes in working practices, and harmonization with staff conditions, also show a tendency to weaken traditional controls over individual payment levels and differentials, and attempts to rupture trade union connections. Esso, the other partner at Mossmoran, had already begun to lay the basis for these new practices at its Fawley plant. In 1983 a Maintenance Enabling Agreement (MEA) was signed, providing for peaks in maintenance workload to be dealt with through a combination of temporary shifts and the increased use of outside contractors. In addition work was reorganized on 'TASK principle', where time, ability and safety became the only restrictions on demarcation. The company also encouraged craft groups to transfer engagements to staff status, which involved a performance-related salary progression. Perhaps as importantly, Esso recognize no unions for staff grades. In 1984 a further measure introduced a level of personal assessment with a personal incremental payment system adding 8 per cent to basic rates over five years (IDS Report 322, 1984).

The Wellcome Foundation, an important pharmaceuticals company, also signed a flexibility agreement in 1986, and in 1987 introduced an individual performance appraisal system establishing a new adult starting rate 3.5 per cent below the grade rate, and two incremental points above it, also of 3.5 per cent. Progression through the increments is subject to performance assessment, and a similar scheme also covers staff grades (IDS Report 498, 1987). For industrial staff performance is assessed on four criteria: job knowledge, work quality, behaviour, including courtesy and cooperation, and attendance and punctuality (*Industrial Relations Review and Report*, August 1987).

More unusually, a new agreement on harmonization at Tioxide, a subsidiary of ICI, led to the loss of recognition for all unions. In this case the issue was improvements to the pension scheme, and negotiations were conducted at plant level by the shop stewards without any reference to union officials, which indicates some weakness in organization, to say the least (IDS Report 498, 1987).

While all these agreements contain some unusual features, or arose, as in the case of Shell, from acute competitive pressures, they contain all aspects of the flexibility question, and are reflected in the general development of collective bargaining in the industry. However, the pace of change has been uneven, and on the whole unions have been able to defend earnings and impose some negotiated limits to the pace of innovation in working arrangements.

In this respect ICI is not unusual. The company began to exert pressure in individual Divisions to integrate some maintenance tasks into process work in 1984, arguing that no new agreements were needed due to the flexibility clauses in the 1969 WSA. The process unions resisted these moves on the grounds that 'The seven-grade structure provided for flexibility only within the traditional categories of chemical, ancillary and maintenance workers [and did not cover] this "new breed of worker"' (*Financial Times*, 17 December 1984). The result was some very hard bargaining, and the union position was well summarized in a T&GWU circular which insisted that

> The Signatory Unions Committee . . . has never taken a Luddite approach . . . to the inevitability of change in respect of the impact of new technology and possible new working methods but insists that the benefits of such change should be shared and that our members should be able to share in the financial rewards resulting. (T&GWU, 1984)

Discussions continued at ICI and the CAIJIC on the pace of change, and the CIA introduced a voluntary craft training scheme aimed at meeting the demands for a 'multi-skilled workforce'. A recent report on changes in the industry as a whole concludes that there have been important changes, but that 'it is unlikely that radical changes in working methods will come cheap' (IDS/PA European Report 293, 1987).

Worker representation in the Federal Republic

The social context of chemical unionism in Germany

The historical and legal framework of industrial relations in West Germany provides a very different social context from that in the UK. In particular the development of codetermination, along with the juridification of collective bargaining, has often been seen as the basis for a more cooperative form of industrial relations centred on the idea of 'social partnership'. These conditions are important in understanding the policy options facing German unions, but 'social partnership' is not an unproblematical concept, and unionism in chemicals shows some distinctive features.

The most striking feature of the evolution of German labour, compared to the UK, is of course the dramatic rupture of fascism. Reconstructing the union movement was not, however, a straightforward process, and was crucially influenced by both internal adjustments to the experience of the Third Reich, and by the policy of the occupying powers.

The allies, particularly the British, regarded the union movement as an instrument for 'democratization', but the US government was suspicious of its political potential. As early as April 1944 the US military government 'envisaged the formation of trade unions, while banning political activity' (Grosser, 1974: 305), and centralized organizations, first on a zonal level, and then in the west as a whole, were resisted until late in 1947 (Schuster, 1973: 74).

The period between 1944 and 1950 not only settled the immediate fate of the 'two Germanies', but also determined the nature of labour's role within the FRG, and its distinctive institutions. At first the unions were active in promoting radical solutions to constitutional and economic questions but their ambitions, centred around codetermination in the economy, quickly took on a conservative hue. The detailed history may be read elsewhere, but for the moment it is sufficient to note three main phases. Firstly the unions looked to the allied 'de-nazification' programme as a means of supplanting the influence of those industrialists most closely associated with the Hitler regime, particularly in heavy industry. The second phase saw some reversals of alliances, as the unions sought a common front with the employers against allied plans to break up parts of German industry, and to restrict certain areas of production. The third phase coincided with the development of the cold war and the establishment of the Federal Republic, and took the form of a struggle over the legal framework of industrial relations, and the form of the economic system.

This evolution had a marked effect on the fortunes of the chemical union, IG Chemie. Established in the Western zone in 1948, the union wanted the industry purged of its association with Fascism, arguing 'that attempts should be made to democratize this powerful force in the West German political economy' (Allen, 1983: 155). This policy was balanced, however, by a wish to maintain the structures of 'Grosschemie' (the term applied to the dominant companies) against allied plans to prohibit or curtail crucial areas of chemicals production. While 'deconstruction' had ceased to be an issue by 1951, the intervening period led to a strategic alliance between the union and the employers.

This alliance did not prove fruitful. IG Chemie's opposition to the breakup of IG Farben was linked directly to its desire to achieve the 'parity codetermination' rights extended to the coal and iron industries by the allies. The partial dismemberment of IG Farben did not eliminate 'Grosschemie', but the union's hopes for extended influence were denied, and the managements of the 'big three' obstructed until the mid-1950s the adoption of labour organization rules, hindering union activity, and contributing to a chronic union weakness in those firms (Markovits, 1986: 290).

The disputes over the legal framework of codetermination which emerged as allied enactments were transformed into federal law were the final phase in the establishment of the German industrial order. Parity codetermination was retained in iron and coal sectors, but the Works Constitution Law of 1951 fell far short of IG Chemie's aspirations. The works councils established under the act had restricted powers, and were to be elected by the employees as a whole, not by union members only. Since many aspects of working conditions were reserved to the councils, and since they were obliged to observe 'social peace', the immediate ability of the unions to organize and represent members at the workplace was severely curtailed. This led to what has been described as a dual system of worker representation, and the union strategy of 'capturing' works councils has played a significant part in the development of industrial relations in the FRG. For IG Chemie, however, problems relating to works councils have taken particular forms that have created recurrent problems for the union.

The employers' organization

Unlike the CIA, German chemical companies split industrial relations from trade matters, and industrial relations are handled by the Federation of Chemical Employers Association (BAVC). Its predecessor, the Arbeitsring, was established in 1949 and continued the 'social paternalism' characteristic of employers in the industry by encouraging 'social partnership' and restricting union influence to a legal minimum. By the BAVC's own account this has led to good relationships with IG Chemie, without requiring a policy of confrontation. However, the real relationships between unions and employers cannot be understood unless the BAVC is seen in the context of the activities of the large firms.

The BAVC is one of the two most important members of the

BDA, just as the VCI is of the BDI, and while formally independent the two organizations have always had close connections which have grown with the passage of time, especially on matters of mutual concern, such as codetermination. The BAVC has often been a vociferous opponent of extensions of codetermination, along with the VCI, which is possibly one reason for the industry's avoidance of state assistance in R&D since 'If they accepted government funding, they would have to allow their union . . . to help plan projects. . . . To avoid strengthening the union's influence, many employers refuse government help' (Allen 1983: 209). The BAVC and the VCI were particularly opposed to proposals in 1980 from the Federal Ministry for Research and Technology, which included mandatory union participation in forward planning, fearing that the state and the unions would 'attempt to extend the concept of Mitbestimmung into the area of industrial research' (Allen, 1983: 210).

The BAVC is a regional organization grouping together regional employers' bodies from twelve areas, but this does not in fact represent the distribution of influence within the organization. On the Standing Committee, which appoints the representatives on both federal and Land bargaining bodies, three regions, Nordrhein-Westfalen, Rheinland-Pfalz and Hessen, hold 55 per cent of the seats. Since these also represent the site of the major works of the three 'IG Farben successor companies' it is plain that influence within the association is quite concentrated, and Ebsworth notes: 'It is no coincidence that, . . . these three regions are the first to negotiate, or that the agreement reached in these regions is adopted almost without exception in the other regions' (Ebsworth, 1980: 90). This concentration of influence has been important in determining the development of collective bargaining, so that 'Far from being a mere umbrella organization, the Arbeitsring is probably the most centralized employers' association in the Federal Republic which has consistently given it a major advantage over the . . . chemical workers' union' (Markovits, 1986: 270). The importance of this concentration is further enhanced by the centralized control over the employers' regional and anti-strike funds by the BAVC, and the continuity of its leading personnel.

In general the BAVC has close relations with IG Chemie, even occasionally attracting criticism from other employers' organizations for being too union-friendly. In 1983 its outgoing President, Bouillon, defended their record, praising IG Chemie as being different from other unions, saying that 'Without doubt IG Chemie adopted a realistic and far sighted policy in recent years, even against internal opposition' (*Handelsblatt*, 13 May 1983). He went on to stress that this closeness did not imply any softness at the bargaining table, and the BAVC continued to secure important agreements on wage structures and the working week, in contrast to employers' experience with IG Metall.

The chemical workers' union

As an industrial union IG Chemie-Papier-Keramik has a wide membership outside chemicals, but as more than half its membership comes from this sector it dominates the union organizationally and politically. Membership density within chemicals is relatively low at about 47 per cent, and this reflects structural peculiarities facing recruitment and representation.

As in the UK, the occupational structure of the industry has diverged from the manufacturing norm, with a relatively large and growing white-collar and technical element and, at the other end of the scale, a higher than average number of unskilled workers. In 1976 white-collar employment was 43 per cent against a manufacturing average of only 28 per cent, and by 1984 had reached over 47 per cent (Ebsworth, 1980: 154; *Handelsblatt*, 12 March 1985). While in-house training has raised the number of workers classified as having some form of skilled status to 65 per cent, the problem of finding 'Common ground between academically trained technicians and the unskilled' (Markovits, 1986: 270) remains a major sources of union weakness.

The problem is accentuated by a further division between the mass of salaried staff (*Angestellte*), and the professional *Leitender Angestellte* which have their own union in the Association of Salaried Academics and Senior Managerial Staff (VAA). The Association grew rapidly from 1960 and represents some 60 per cent of the relevant grades. There has also been a noticeable increase in the proportion of white-collar workers in higher-paid grades, with 38 per cent of commercial staff and 43 per cent of technical staff in the top two grades. Salaried membership of IG Chemie, however, tends to be concentrated at the level of foreman or lower-grade technician, possibly reflecting their closeness to the shop floor. The BAVC found up to 95 per cent union membership amongst new entrants in production, but only 13 per cent in laboratories and sales staff (*VB Finanzen-Angestellte*, 1986). This factor has 'contributed to an internal split which has perpetuated a "white collar" mentality (*Angestelltenmentalität*), and has hindered IG Chemie's potential for mobilization' (Markovits, 1986: 272).

A second problem for IG Chemie lies in its relative weakness in Grosschemie, most especially in Hoechst and Bayer, two of the IG

Farben 'successor companies'. This is again particularly noticeable in the white-collar grades, with membership figures of some 18 and 15 per cent at Bayer and Hoechst, but over 38 per cent at BASF. This reflects management styles in the companies, preceding even the formation of IG Farben, which have been notable for both their 'paternalism' and anti-union sentiment.

A survey of Bayer in 1986 illustrates the continuity of this policy. For example, the company employs fifty-nine qualified social workers and four industrial clinical psychologists. Involvement in the lives of the workforce extends even to sport, with two major football teams being sponsored, and Bayer athletes taking twenty-three Olympic medals over the years (*Management Wissen*, April 1986). This willingness of companies to 'buy off' union members is closely connected with the role of the works council, which is discussed below.

Both law and employer strategies have limited the direct role of the unions within companies, which has lent a centralizing tendency to their organization, a trend which is somewhat more noticeable within IG Chemie. The union shares the organizational forms common to most German unions, with a central administration overseeing regional committees which are, in principle, the site for collective bargaining. In addition there exist district structures, or branches, which are 'often described as the backbone of all union activities and the main locus of IG Chemie's organizational power' (Markovits, 1986: 278). Nevertheless effective power within the union is concentrated among the full-time officials at the centre, on the one hand, and members of the major works councils on the other.

Ebsworth has described the practical concentration of power in the hands of the National Executive Officers, in the formulation of collective bargaining aims, and control of joint conciliation procedures. Much of this centralization of power can be traced to the growing coordination amongst the employers, and the influence within the BAVC of the 'big three' (Ebsworth, 1980: 66). In addition the national officials also exercise considerable patronage throughout the union. All eight Regional Secretaries are appointed by head office and 'As a consequence, most of the District Regional Secretaries seek to please the executive committee, their only official source of legitimation, rather than the regular union members' (Markovits, 1986: 275). This combination of centralization and patronage should not be overemphasized, but it is more marked in IG Chemie than in other major German unions.

Dual representation and IG Chemie

In German law collective bargaining over wages is formally reserved to the trade unions, but many issues concerning the organization and conditions of work are in the hands of the works councils. The unions have responded to this duality by running lists of union activists in elections, and in many cases have succeeded in 'capturing' the councils for the union, so that they represent 'the extended arm of the union under codetermination' (Streeck, 1985: 12). In the case of the car industry Streeck has argued that strong union organization has not only transcended the dualism of the system, but has influenced corporate policy towards production strategies based on employment stability and the utilization of high skill level (Streeck, 1985: 12–21). An important element of union influence is found in the development of the *Vertrauensleute* system of in-plant organization pioneered by IG Metall. The *Vertrauensleute* (literally 'trusted people') do not function as shop stewards in the British fashion, but they do represent the organized presence of the union within the workplace, the source of union candidates for the works councils, and an important link between the union's officials and the membership.

In respect of IG Chemie this general picture must be modified, and tensions between works councillors and the *Vertrauensleute* have been an important factor in the internal politics of the union. Following the recession and the wave of (often unofficial) strikes at the end of the 1960s the *Vertrauensleute* system became the focus for left-inclined unionists wishing to expand the part played by the union in the workplace. In chemicals too this was associated with the 'Activist section of the union', who were 'collectivist, reform minded, and linked to a relatively radical tradition in the German labour movement' (Markovits, 1986: 294).

The *Vertrauensleute* play a unique role within IG Chemie with the exclusive right to elect the delegate council of the branch, the prerogative of the whole membership in other unions. As a result 'The composition of the union's congress, its plenum and ultimately its key committees . . . originate with the shop stewards at the plant level and in the locals (branches)' (Markovits, 1986: 279). However, what seems on the surface a powerful basis for union organization has, in fact, proved a powerful source of dissension. The employers responded early to the threat of an increased union presence within the workplace by organizing elections amongst all employees for 'company shop stewards', and by the end of the 1970s forty-two major plants had committees of these representatives in place. A practice grew in the union of recognizing as *Vertrauensleute* anyone elected in this fashion, provided they were at least union members, and extending that status to all works councillors. This has meant that 'The presence of these "plant appointed" . . . shop stewards . . . within the union's organizational domain has meant that non-unionized chemical workers have exerted a direct influence upon some of the most crucial decision making bodies within IG Chemie' (Markovits, 1986: 280).

An attempt by leading works councillors and their supporters on the National Executive Committee to write this arrangement into the rule book at the 1980 congress led to a major dispute, with accusations of an 'authoritarian leadership style' before a compromise was reached subjecting 'plant-appointed stewards' to a separate election by union members (Allen, 1983: 207).

The dispute underlines the particular difficulties facing the union in terms of the Works Constitution Act, and the role of the works councillors in Grosschemie. DGB unions have proved very effective at 'capturing' the works council structures, and IG Chemie is no exception, with over 90 per cent of workers' representatives, and over 85 per cent overall (Ebsworth, 1980, II: 112). It is also true, however, that this success is associated with the larger plants, and that it has to a degree resulted in works councillors 'capturing' the union. This is a matter of degree, but in effect the role of the big three, in weakening the union within the plants, has been also to weaken it within the industry as a whole. Opposition to these trends led to the expulsion of 275 members for supporting opposition lists in works council elections between 1972 and 1975 (Miller, 1982: 60; Ebsworth, 1980: 306n).

The influence of the union was extended by the Codetermination Act of 1976, which established worker representation at company level in public companies with more than 2000 employees by granting parity representation on supervisory boards, and a duty on that board to appoint a Labour Director to the management board. The act was fiercely opposed by the employers in general, but particularly by the chemical industry which, by virtue of its structure, accounts for 10 per cent of the 600 companies likely to be affected. An appeal by employers to the constitutional court seriously delayed the act, causing the DGB unions, including IG Chemie, to withdraw from Konzierte Aktion, the tripartite economic policy arrangement.

In fact the Act, far from leading to the 'socialist workers state' allegedly feared by the Chair of the Executive Board of Bayer, was a limited and partial step towards the union aim of 'parity codetermination'. Senior management gained the right to separate representation as 'employees', and there is no veto on the appointment of the Labour Director (Ebsworth, 1980: 111–12). In the first elections under the act 71 per cent of all seats were won by DGB unions in the industry (or 83 per cent of available seats, if management personnel places are discounted). While this is a respectable result, in line with the policy of 'capturing' works councils, 'Comparison with the overall position within the organizational area of the Chemical Workers' Union shows that it is below average, reflecting perhaps the [union's] weakness . . . in some of the larger chemical companies' (Ebsworth, 1980: 120).

Collective bargaining in chemicals

Collective agreements operate at two levels in the chemical industry. Every four years a framework agreement (*Manteltarif Vertrag*) is concluded at the federal level setting hours of work, holidays, and retirement ages as minima for the entire industry. Up until 1987 a further framework agreement was negotiated at regional level to establish the relationship between the four wage groups – manual, white-collar, technical, and academic. Actual wage agreements are determined on a regional basis annually. In principle these agreements are determined autonomously in the eight regions, the union demands being formulated by a bargaining committee composed in its majority of workers rather than officials. In practice the process is more complicated, and the autonomy of the regions is circumscribed in two important ways.

Firstly, contracts in the three regions with the highest concentration of companies and plants (Rheinland Pfalz, Hesse and Nordrhein Westphalia), which includes the union stronghold amongst the big three at Leverkusen, expire a month earlier than other regions. Negotiations in these regions are carefully coordinated, making them 'IG Chemie's "vanguard" whose settlements usually are transferred to the rest . . . of industry' (Markovits, 1986: 278).

Secondly, these centralizing tendencies are reinforced by the role of officials at the centre. At the start of each bargaining round a meeting of the National Executive and the Regional Secretaries establishes bargaining guidelines. If a region chooses to exceed the guidelines, or proceeds towards a dispute, the union cannot interfere. However, since by law and by agreement conciliation must precede a dispute, the union centre has another chance to determine the outcome. At conciliation meetings only one of the three union representatives is from the affected area, and as the NEC has a representative, the centre can choose to use its vote as an effective veto on action by voting with the employers' side as occurred in the Nordrhein in 1977 (Ebsworth, 1980: 203, 251).

Formally speaking the regional agreements should be the end of the matter, but in fact a second round of bargaining takes place at plant level, under the auspices of the works councils. Since it is illegal for the councils to negotiate over wages, or to reach a collective agreement, they use matters subject to codetermination, such as overtime, as a form of 'sub rosa' bargaining. This secondary bargaining has assumed great importance in the chemical industry, especially in the big three. Actual differences are hard to determine, varying from company to company, and across subsectors, but appear highest in the large companies where second-round bargaining can add between 12 and 20 per cent to the official wage levels. A survey of Bayer in 1986 illustrates the importance and extent of these benefits. For every Deutschemark spent on wages and salaries the firm spent DM1.07 on social and fringe benefits, including pensions, housing, social and family assistance, as well as loyalty payments (*Management Wissen*, April 1986).

Streeck has argued that this activity is acceptable to the unions since industry-wide bargaining systems 'Require a correction mechanism at the level of the individual firm which is strong enough to make the central agreement generally acceptable, and weak enough not to render it . . . meaningless' (Streeck, 1981: 159). In the case of chemicals, however, the function of such bargaining is dictated by the constraints on the union in the big three which has meant, in some degree, 'ceding power to the fiefdoms of the works councils "princes" (*Betriebsratsfürsten*)' (Markovits, 1986: 283), in addition to the privileged place they occupy within the formal bargaining system.

The union was very successful in establishing wage and salary levels between 4 and 8 per cent above the industrial average throughout the 1960s and 1970s, although this success was less for salaries in the latter part of the period (Ebsworth, 1980: 186–7). This success was due to several factors, including high growth rates, and the policy of 'following' IG Metall, as well as a conscious decision by the leadership to press the wages question following their disappointment over the extension of codetermination in the 1950s. While acceding to wage demands, however, the employers were determined that 'Under no circumstances was IG Chemie to be allowed to translate its . . . bargaining power into an increase in shop floor authority and plant level control' (Markovits, 1986: 297). One aspect of the employers' response was to increase their own coordination, developing a 'Tabu Katalog' of forbidden concessions backed by heavy sanctions and fines. By 1969 this pressure had led to a marked increase in centralization in bargaining, and considerable dissatisfaction among the union activists, which the leadership countered with references to 'the increasingly politicized nature of collective bargaining rounds and their link to the economic situation of the country as a whole' (Markovits, 1986: 297). This dissatisfaction opened a new period in collective bargaining strategy which produced significant conflict both with the employers and within the union, and which did much to weaken the union's response to the problems of the late 1970s.

The union opened a campaign in 1970 based on an aggressive return to regional autonomy. This policy was resisted not only by the employers, but also by the works council at Hoechst which argued explicitly 'that it would make their own efforts to obtain benefits from the company redundant' (Markovits, 1986: 303). The result was something of a mixture, because although eventual settlements were the highest in the history of the Federal Republic, the union experienced severe difficulties in coordinating actions, most particularly in the case of their only stronghold in the 'big three', BASF. The most marked success was in Hamburg, which experienced the first chemicals strikes since Weimar, and which won a 'thirteenth month' bonus which strengthened the morale of the left wing of the union (Markovits, 1986: 303).

The next round in 1970 proved far more difficult for the union. The industry had experienced a severe recession in 1970, and the employers were committed to a more determined resistance, establishing a 'Fund for the Securing of Labour Peace', to which each company had to pay 0.7 per cent of its 1969 wages bill (Markovits, 1986: 304). Nevertheless the union conducted a vigorous thirty-day campaign, involving thousands in various forms of industrial action and demonstrations while encountering problems of violent attacks from strike-breakers, clashes with the police, and actions against them in the courts (Markovits, 1986: 308).

Despite a relative success this wage round marked a watershed in the union's bargaining strategy, although this was not to be immediately apparent. By 1974, however, divisions within the union, concentrated around the behaviour of the 'big three', coinciding with a general demobilization in the leading sector of engineering, allowed the employers to restrict wage advances, while at the same time structural demands from the union remained relatively underdeveloped.

Important as were these developments in the 1970s they did not imply any fundamental break with the 'industry-friendly' nature of the union. It reacted sceptically to the DGB on 'qualitative bargaining' during the recession, and distanced itself from the campaign for a shorter working week spearheaded by the metalworkers. The union view is that its realism maintained wage settlements in a period when time reductions were unlikely to be gained.

As recovery proceeded changes in production and the utilization and classification of labour became the object of extended negotiations between the union and employers. It is noticeable that these negotiations have followed a very different path from the changes in the British industry. In 1987 a new 'framework agreement' was signed which, for the first time in a major industrial sector, established a single pay scale for all employees, and which initiated the movement towards a shorter working week.

The agreement, which took six years to complete, and which had to go to the industry's arbitration body, will last for seven years, and sets basic salary improvements for the first three years. The new thirteen-grade salary system is designed to ensure equal pay for work of equal value, and IG Chemie 'has expressed satisfaction that in the areas where skilled workers and craftsmen overlap with clerical and technical employees there will be much greater parity between the salaries paid' (*European Industrial Relations Review* (*EIRR*), Report 164, September 1987). The changes took place over three years, with full protection for those grades, especially in technical areas, where a reduction in salary would otherwise be necessary. In the third year the basic week was reduced to thirty-nine hours, with reduction to thirty-five hours for workers over fifty-eight years old. There will be no further changes in working time before 1992.

Detailed changes in job descriptions, and arrangements for the shorter working week, will need to be negotiated by works councils. Guidelines have been issued which emphasize that the valuation of a job will be determined solely by the work performed, and the level of skill and responsibility involved. The chief executive of the employers' federation laid great stress on the unitary nature of the agreement: 'The social significance of the agreement cannot be overestimated. Obsolete and unjustified differences in status have given way to a simple and workable pay system for all employees' (*EIRR*, Report 164, 1987: 10).

Unions and public policy in the UK and the Federal Republic

In general there are two levels at which unions represent their members in matters of public policy. The first is at the 'macro' level of government policy relating to economic and social policy, wage determination and the conduct of industrial relations. The second (but interdependent) level is that of industry or sector-specific questions, from 'external' questions such as trade or competition policy, to 'internal' issues concerning the application of specific sets of regulations such as health and safety. Given the connections between the levels it may be expected that industrial unions such as IG Chemie will judge issues more explicitly from a sectoral perspective. There is a deal of truth in this, and IG Chemie's greater association with its industry has already been noted on the question of responses to the 'new producers'. However, even industrial unions operate within the general framework of a national labour movement, and under non-sectoral political constraints.

The connections between industrial and political concerns are recognized by the German unions for whom 'Generally speaking industrial relations are a device designed to bring about a transfer of bargaining power from the sectional to the societal level and to enable the union movement to pursue macro-economic and macro-political objectives' (Streeck, 1981: 153). While such an explicit formulation would be more controversial amongst many British unions the general tendency in the UK (at least until 1979) has been for the unions to aspire to the representation of all working people at a political level, a view developed in the heyday of tripartism in the 1960s and 1970s. While it is not possible here to develop this area extensively, some distinguishing features of the place of chemical unions in the public policy arena can be outlined, and specific attention is paid to the regulation of health and safety issues.

In contrast to Germany, union involvement in public policy-making has been deeply influenced by the 'voluntarist' tradition in collective bargaining. In debates over industrial democracy the chemical unions have had deep reservations about the effectiveness of board-level representation, and preferred to concentrate on the extension of information rights and the development of collective bargaining to embrace a sectoral forum which could link up with government economic policy. This attitude also reflects a greater scepticism in respect of the commitment of the employers to the UK industry than is the case with IG Chemie.

Consequently the unions put a deal of effort into playing a part in the sectoral work of the NEDC, and the many joint bodies that provided the linkage with government policy. However, since this policy relied on an interventionist form of government for its success the opportunities for union influence have been severely curtailed under the neo-liberal policies of the Conservative government. A case in point is the abolition of the tripartite Industry Training Board for the industry, which not only oversaw the financing of training, but established some important standards for skill definition and development. The effective exclusion of the unions from centralized decision-making on training has been seen as particularly damaging in the context of rapid changes in work organization and the introduction of skill flexibility.

In Germany the general context is very different, not only because of tighter legal definitions of union rights, but because of the rights available under codetermination. Despite the period of Konzierte Aktion the articulation of wage determination and government economic policy never produced the extensive tripartite structures which characterized the UK in recent years. Instead this articulation was more dependent on the role played by union influence within federal and Land Labour Ministries, which is further discussed below. This is not to say that industrial relations was not highly politicized. During the unusual period of wage conflict in 1970 the IG Chemie leader Hauenschild stated that 'I have the feeling that the employers . . . are trying to provoke a showdown. . . . They beat up on the chemical workers, but they really mean to beat up on Bonn' (Markovits, 1986: 307).

The experience of tripartism in the UK, and its subsequent decline, was not a great success for the unions. At the height of their influence under the Industrial Strategy of the 1974–79 Labour administration the unions were able to secure government backing for safeguards for employment and recognition when the American multinational Tenneco took over Albright and Wilson, a major UK company. Assurances were given that 'Tenneco will not interfere with the government's industrial strategy in which Albright and Wilson through its involvement in the EDC is important, and that the industrial relations policy of [the company] would be left intact' (Prest, 1978).

At a more fundamental level, however, they were unable to secure their strategic aim of a rapid expansion of plastics capacity through the development of an integrated gas-gathering system to exploit North Sea feedstocks, despite formal agreement at the Chemicals Economic Development Committee. The failure of that initiative, and subsequent changes in government policy, have reduced the EDC to a 'talking shop', and underlined the inability of the unions to challenge what they characterize as an irresponsible industrial approach to domestic investment and job security.

In Germany the IG Chemie has long held a more 'industry-friendly' approach which, in common with most other unions, has been expressed as a 'De Facto "growth alliance"' (Jacobi, 1983: 1) with the employers. This has been particularly noticeable in the union's defence of the industry against criticisms from environmentalists, which extends as far as organizing internal surveillance of its own 'chaoten' elements. The essentially conservative outlook of the union has been emphasized by its policy towards the use of Branchenfonds schemes to develop worker shareholdings. In contrast to the majority policy of the DGB to develop these funds on a sectoral basis as a means of influencing major investment decisions, IG Chemie has effectively accepted company-based funds. As a result the industry now dominates the 300 'partnership firms' in the Federal Republic, and this has been seen as 'An expression of the overall shift towards an accommodationist posture which IG Chemie experienced after the onset of the crisis in 1975' (Markovits, 1986: 317). This disagreement emphasizes the fact that the union is often seen as being 'out of step' with its partners in the DGB where it is an important element on the right. Along with the miners, construction workers, textile and food-processing industries it is part of the 'gang of five' which has opposed the more aggressive policies associated with the metal-workers and printers who head the majority of DGB unions.

The regulation of health and safety

The regulation of occupational health and safety is a major area where the collective bargaining role of trade unions intersects with public policy and government administration. While conventional wisdom maintains that these matters ought to be of equal and mutual concern to employers and workers (a concept enshrined, for instance, in the tripartite constitution of health and safety regulation in the UK) the promotion of a safe work environment has been marked by deep conflicts of interest. Even a cursory reading of the history of the subject reveals the extent to which employers have been able to harness governments, and the medical and scientific establishments, in defence of short-term production interests as against the protection of workers. A classic example may be found in the autobiography of the British miners' leader Arthur Horner (1960: 141–6) concerning the establishment of silicosis as an industrial disease, while a more recent example is the long period taken to establish ICI's liability for Nonox S, responsible for cancer of the bladder in rubber workers (Kinnersly, 1974: 129–30). It should be remembered, too, that many health issues concerning working practices, especially those relating to stress, intensity of work, and the effects of shift working, are often dealt with outside the formal regulatory procedures, and are 'disguised' as purely industrial matters. A recent report by the ETUI, for example, stresses union concerns over 'hidden' repercussions of the drive for extended work flexibility, and other aspects of deregulation of employment conditions (IDS/PA European Report 293, 1987).

While the chemicals industry does not spring to mind as an obviously dangerous occupation in the same way as mining or construction, the potential hazards of a major plant failure are obviously immense both for workers and the general public, as has been seen in the disastrous accidents at Flixborough in the UK, Seveso in Italy, and Bhopal in India. At a more mundane level chemicals share all the hazards of any major industry engaged in continuous process production and, perhaps more importantly, in both the production and consumption of its products exposes workers to chronic as well as acute threats from chemical substances many of whose effects are poorly understood or simply unknown. Against this background a comparison of the role played by unions in Germany and the UK is an important indicator of the effectiveness of worker interest representation, and one which shows marked differences in power and the attitudes and behaviour of employers.

The Federal Republic

The regulation of health and safety in the Federal Republic is shaped by the federal nature of the political system, the place occupied within the state by the unions, and specific developments in the civil service ethos. The principal government agency involved is the Federal Ministry for Labour and Social Affairs, which is responsible for developing general policies on worker protection at the workplace. One of its most important functions is, in conjunction with the Deutsche Forschungsgemeinschaft, to set the so-called MAK standards for maximum allowable exposure to hazardous substances. The development of detailed regulations is the task of the Federal Institute for Occupational Safety and Accident Research (Bundesanstalt für Arbeitsschutz und Unfall-Forschung), which is attached to the federal ministry. Inspection and enforcement is the responsibility of the Staatliche Gewerbeaufsicht, whose factory inspectors (*Gewerbeaufsichtsbeamten*) are part of the labour or social ministries of the Länder governments. This is in line with the general administrative principle that Bonn ministries have no field services.

Labour ministries are almost invariably held by trades unionists, whether at Land or federal level, and this practice even extends to the CDU, where the minister, as with Norbert Blum in 1987, is usually a member of the Catholic trade union wing of the party. This trade union orientation is not limited to the minister alone. It is a striking feature of the West German political system that the 'Chinese Wall' between politics and administration has collapsed, allowing senior public servants to play an openly partisan role

(Grant et al., 1988: 276). Thus the labour ministries are staffed largely by people sympathetic to the unions. Additionally the shadow Bundestag committee monitoring the work of the ministry is dominated by members with a trade union background. In total this ensures the unions a very high degree of success and influence in the state regulation of health and safety.

There is a second layer of regulation embodied in the Berufsgenossenschaften which originated as a system of industrial accident insurance, and whose core task remains accident prevention. Organized in thirty-five industrial sectors these bodies reflect the high priority given to *Selbstverwaltung*, or self-administration, in German administrative practice. The Berufsgenossenschaft Chemie, founded in 1885, is based in Heidelberg and has about sixty *Technische Aufsichtsbeamten* (TABs, or technical inspectors), to monitor safety standards in chemical plants.

The Berufsgenossenschaft (BG) has a quasi-state character with the power, originating in a law of 1911, to issue binding rules and to impose sanctions for breaches, although regulations require the approval of the Labour Ministry. They must now be approved also by the EC Commission as being compatible with competition laws. The dualism of the BG system is seen as creating some problems for coherent regulation, and the distinction between it and the official state system is to be reduced in the interests of 'user-friendliness'. It is already the case that the two inspectorates check that each other's regulations are being observed, and both are involved in the planning of every new plant.

There is a central body of the BGs in Bonn with twenty-six specialist committees which liaise with the Labour Ministry, and BG Chemie is responsible for the Committee on Chemicals and Pressure Vessels. Very active in the area of chemicals control, the BG Chemie has two main approaches to the control of hazardous substances. The first is a test programme for substances with long-term effects, and is the major programme in the Federal Republic for dealing with the *Altstoffproblematik* – the problem of existing chemicals. The second approach involved the establishment of a test programme for possible carcinogens in 1978. Run by an advisory body drawn from industry and the sciences it refers suspect substances to the Deutsche Forschungsgemeinschaft for the establishment of MAK values.

Both sides of the industry are represented on the BG Chemie on a parity basis. The majority union presence is from the IG Chemie, but the minority unions for staff and academic employees are also represented. The employers' nominees are appointed by the BAVC. Two chairpersons are elected, one from each side, who rotate half-way through a six-year period. Both elements of this dualistic system of regulation provide a very high level of access and influence to the unions. In addition the high levels of technical expertise of the industry, and its strong financial position, mean that high union expectations on safety standards are usually met.

The United Kingdom

State regulation of safety at work has a long history in the UK, but its development was very fragmented involving numerous Acts of Parliament, and often glaring omissions in coverage. The current system arose from a report of a committee under Lord Robens, established by the Heath government and published in 1972. The tripartite approach developed was a part of an attempted accommodation with the unions towards the end of the Heath government, and the report resulted in the Health and Safety at Work Act of 1974 (H&SW Act), which is essentially an enabling act, specific regulations being promulgated on recommendations from the Health and Safety Commission. The new legislation covered most of the committee's recommendations, with the effective exception of those on the monitoring and protection of occupational health, which has become an area of increasing concern to the unions.

Nonetheless the act had considerable support from the unions, both as a systemization of safety law and as an aid to union organization and access to policy-making. The central policy body, the Health and Safety Commission, is organized on a tripartite basis with union membership nominated by TUC, and is serviced by an extensive network of tripartite consultative committees covering individual industries. As importantly the act gave legal rights to trade unions to appoint safety representatives on the shop floor, and to require the establishment of joint safety committees in firms of over 250 employees.

The Commission consists of three representatives from the employers (in practice nominated by the CBI), three from the unions and two from local authorities (one from England and Wales, and one from Scotland). The Commission is responsible for the formulation of regulations, which must be approved by the Secretary of State for Employment. Along with its executive arm, the Health and Safety Executive (HSE), the Commission is thus wholly responsible for developing and enforcing policy in this area. The Commission, however, is not in a position to completely dominate the HSE since:

> Because the HSC by law cannot intervene in specific cases, the HSE has scope for exercising discretion. . . . The HSE, moreover, has greater practical and technical knowledge than the part time members of the

> HSC or the pressure groups they represent; the HSC . . . is therefore unlikely to be able to refute technical arguments from the HSE. (Wilson, cited in Grant et al., 1988: 279)

To assist the collection of industry-specific data, and to carry out negotiations with interested parties, twenty-two National Industry Groups (NIGs) were set up within the HSE in 1977. The National Industry Group for Chemicals is located at the HSE headquarters in Bootle, and its major function is to produce reference materials for the factory inspectorate. It has produced a two-volume *Guide to Inspection at Chemical Works*, and issues *NIG Minutes* on particular topics, and arranges training courses for factory inspectors. It is noticeable that the resources of the British system are far more limited than those of its West German counterpart. While the factory inspectors are generalists, as are their German equivalents, the specialist input of the NIG is very limited compared to that provided by the technical inspectors of the BG Chemie. Every one of the TABs is a graduate chemist with 4 years practical experience and 2–2½ years specialist training. Compared to this the sectoral resources of the HSE look far from adequate.

Fear of major plant failures, like the Flixborough explosion of 1974 which killed twenty-nine, led to regulations on the Control of Industrial Major Accidents and Hazards (CIMAH) which were applied in the UK in stages between 1984 and 1989. In the chemical industry the NIG established a joint working party with the CIA to control the bulk storage of chlorine. This working party was very unusual in that it had trade union representation throughout its working life. The more typical pattern is for unions to be consulted at a very late stage. This reflects a marked lack of enthusiasm on the part of employers for trade union involvement, and is a feature which distinguishes chemicals from many industries in the UK, and from the industry in Germany.

Multi-unionism, and the general rather than industrial nature of the unions concerned, is undoubtedly a handicap in overcoming employer hostility to participation in health and safety regulation. There is little possibility of their developing a specialist section for environment and health and safety matters, as IG Chemie has done. It may also be argued that multi-unionism has militated against the very close relations between employers and industry typical in the German industry. Union support in resisting external attacks on the German industry recently has encouraged chemical employers to develop union participation in health and safety regulation. However, such a close relationship clearly involves dangers of an overidentification of interests that could undermine union effectiveness. More importance may be attached to the political position of the German union, being able to draw on a tremendous fund of expertise and friendly officials in the Labour Ministry, and from its participation in BG Chemie.

Of far greater importance than multi-unionism is employer hostility to the broadening of union influence, and the politically unstable environment of health and safety regulation in the UK under the present administration. There has been a long-running controversy over the failure of the HSE to establish an Industrial Advisory Committee for Chemicals, which was consistently opposed by the employers who argued that the unions lacked the necessary expertise. A compromise was reached with the establishment of a National Joint Advisory Committee on Health and Safety in the Chemical Industry under the auspices of the CIA. This committee has been generally welcomed by the unions, but criticisms remain. The employers have resisted the appointment of the GMB's national safety officer (then one of the few professionally qualified officials among the unions) to the committee on the grounds that he is not a national official party to industry negotiations.

The major unions have produced critical reports of health and safety within the industry, which stand in sharp contrast to the attitude of IG Chemie, which invariably praises its industry's record. The GMB issued two major documents on health and safety in chemicals in 1985 and 1987, which argue that the industry's record is poor by manufacturing standards in general, and that it has been deteriorating in the recession, an accusation denied by the CIA. In particular the union accuses the chemical and pharmaceutical industries of delaying the introduction of the Control of Substances Hazardous to Health regulations which are due to come into force in 1989 (Grant et al., 1988: 281; GMB, 1987: 9).

As a further demand on the industry to improve controls and take account of the interests of its employees, the 1987 national wage claim included proposals for a Work Environment Fund, financed by a 0.15 per cent levy on all employers, to be administered jointly by employers and unions. The idea is modelled on the Swedish system, established in 1972, which currently spends the equivalent of some £50 million a year on research and its application, training, information on codetermination and assistance to safety delegates. The GMB points to the lack of resources for research in the UK, and the virtual exclusion of unions from influence in the area. While little may be expected from industry negotiations, the union is pursuing the idea with the Labour Party in the hope of influencing a public debate (GMB, 1987: appendix 12).

The unions have also been highly critical of the restricted funds available to the HSE, and of the decline of the inspectorate which has resulted. In particular they were alarmed by the HSC's Plan of Work produced in 1986, which proposed a radical shift to industrial self-regulation involving extra duties for local authorities, and the possibility that insurance companies would be allowed to take over inspection in some industries. The proposals were eventually withdrawn for reconsideration, being also opposed by the employers' representatives. Fears remain, however, that government commitment to deregulation, coupled with its restrictive funding policy, will seriously undermine health and safety regulation just at a time when more resources are required for the expansion of occupational health monitoring.

The controversies over the future of the HSC illustrate the important differences between the chemicals health and safety area in the UK and Germany. In Britain companies dominate the area far more, and they possess overwhelming advantages in relation to the information and influence they can bring to bear on policy information. The resources of the unions, and even of the HSE, hardly bear comparison with the impressive technical and political resources enjoyed by IG Chemie and the BG Chemie. The marked degree of trust that exists in the German case is clearly missing in the UK, and while health and safety is largely conducted on a consensual basis the area of workers' protection is more adversarial in the UK.

This adversarial relationship is not peculiar to chemicals, but reflects a broader unwillingness amongst British employers to increase the regulatory role of unions. The CBI has recently voiced fears over nine new health and safety directives being considered by the EC, and amendments now before the European Parliament on the control of benzene. The proposals, which are said to draw heavily on Scandinavian practices, would extend the element of codetermination in the health field, which the CBI regard as being 'social engineering' by the back door. The International Health and Safety Advisory Group of the CBI has been strengthened to alert British industry to the possibility that the proposals would work against the 'change in the balance of industrial power over the last eight years' (*Daily Telegraph*, 1 December 1987).

Conclusions

Union organization in the British chemical industry developed strongly after the war and, after initial difficulties, made substantial inroads into the growing white-collar occupations. This growth was strongly linked to the general development of the labour movement, rather than to sectoral factors. All the unions are general rather than industrial, but their chemicals membership forms a distinctive sector with a stable and coherent bargaining structure. The major unions are all prominent in labour and union politics, and have been closely associated with policies linking collective bargaining with sectoral planning. Unlike IG Chemie, worker representation in the UK industry is therefore influenced by a broader conception of industrial interests.

Collective bargaining in the industry has nevertheless developed its own character, being relatively conflict-free, and often innovative in respect of productivity. The comparative rarity of conflict does not indicate union weakness, or an uncritical acceptance of management views of the industry's interests. The unions pursue their own views on the future of the industry, and these have at times diverged sharply from corporate and government views. Stability in the industry has owed more to a good growth record providing opportunities for negotiated change than to any cooperative culture specific to the sector. One official summed up the position by saying 'The employers do not really want trade unions, but while we are here they will deal with us intelligently.' There are signs, however, that employers are increasingly willing to challenge or evade union organization despite the sector's industrial relations record.

Labour organization in the German industry is very much the child of the postwar order, despite legal continuities from the Weimar period. The formal divorce between the unions and the works council organization, which has done much in shaping the union movement as a whole, has had a particular effect in chemicals, weakening union organization and distorting its internal politics. This is linked to another important continuity in the dominant role played by the 'big three' successor companies to IG Farben which combine paternalism with hostility to union influence. This has contributed to the union's weakness in the growing white-collar sector, although this may possibly be modified by the new agreement on a unitary pay structure. Despite its presence in other industries the chemical sector is the union's political and organizational heart. Closely associated with the SPD, and the third largest union in the DGB, IG Chemie is nonetheless primarily oriented to its chemical interests, and its political attachments are not allowed to interfere with its wages policy. Although far from being unique in its sectoral orientation IG Chemie is nonetheless markedly 'industry-friendly' in German terms, often defending 'its' industry against critics in the labour movement, and cooperating closely with employers on environmental issues. This cooperative attitude has extended to investment policy where IG Chemie is far more supportive of the industry's internationalization than the British unions.

The combination of codetermination and a relatively low level of government intervention has produced a very different environment for trade union influence in Germany than in the UK. In the UK the unions were closely identified with tripartite planning as a vehicle for developing worker influence and the scope of collective bargaining. The experience of tripartism has, however, been a contradictory one for the unions. For a time it did provide them with a degree of legitimation for their claim to influence the general development of the industry, as well as a platform from which to criticize the influence of both government and corporate policy. However, the consensual nature of tripartism at the same time excluded radical challenges to corporate power. In the political climate of the 1980s, moreover, tripartism, even where its institutions have survived, is increasingly empty of any real content.

Tripartism of this sort has been totally absent in the German context, and it is difficult to imagine either IG Chemie or the employers welcoming it. In part this is due to the much stronger identification of the union with the industry, and in part to the possibilities under codetermination. Yet it is far from clear that codetermination has provided the union with much of a lever on the major developments within the industry. If consensus is the basic assumption of tripartism in the UK, in Germany it has a legal basis in the institutions of codetermination, and the commitment to social peace has marked all levels of industrial relations. The particular influence of the leaders of the major works councils within the union has often been more a matter of controversy than of strength.

Nevertheless IG Chemie enjoys some marked advantages over the British unions in many areas which arise from the legal status of collective bargaining and codetermination, and from its avenues into Labour Ministries. These differences are clearly seen in the part played by the union in the regulation of health and safety matters in the industry. In the UK, despite important advances under the terms of the H&SW Act, the unions have found it more difficult to develop a proactive safety policy, and they face constraints from both institutional limitations and employer hostility. For British unions, therefore, safety matters are seen as being closely linked to the development of trade union organization at plant level, to a more marked degree than in Germany.

Despite the differences between the two countries it is still plain that they share many of the same inequalities of power in respect of the employers. While they have, for the most part, clear avenues of influence in basic labour matters, and have established the industry as relatively well paid and stable, the employers have expended considerable efforts to restrict union influence to as narrow a scope as possible.

References

Allen, C.S. (1983) 'Structural and technological change in West Germany: employer and trade union responses in the chemical and automobile industries'. Unpublished doctoral dissertation, Brandeis University.

Bain, G. (1967) *Trade Growth and Recognition*. London: HMSO.

Bain, G. and Price, R. (1983) 'Union growth: dimensions, determinants and destiny', in G. Bain (ed.), *Industrial Relations in Britain*. Oxford: Blackwell.

Brown, W. (ed.) (1981) *The Changing Contours of British Industrial Relations*. Oxford: Blackwell.

Coldrick, A. and Jones, P. (1979) *International Directory of the Trade Union Movement*. London: Macmillan.

Eaton, J. and Gill, C. (1978) *The Trade Union Directory*. London: Pluto.

Ebsworth, D.R. (1980) 'Industrial relations in the West German and British chemical industries'. Unpublished doctoral dissertation, University of Surrey.

Elias, P. (1985) 'Changes in occupational structure, 1971–1981'. Unpublished paper presented to the Edinburgh Survey Methodology Group.

Gill, C., Morris, R, and Eaton, J. (1978) *Industrial Relations in the Chemical Industry*. Farnborough: Saxon House.

GMB (General, Municipal, Boilermakers and Allied Trades Union) (1987) *Health and Safety in Chemical and Allied Industries*. Surrey: Municipal Boilermakers.

Grosser, A. (1974) *Germany In Our Time: A Political History of the Post War Years* (translated by P. Stephenson). Harmondsworth: Pelican.

Grant, W. (1983) *The Organization of Business Interests in the UK Chemical Industry*. Berlin: Wissenschaftzentrum.

Grant, W., Paterson, W. and Whitston, C. (1988) *International Industry, National Governments, and the European Community: A Comparative Study of the Chemical Industry in Britain and West Germany*. Oxford: Oxford University Press.

Horner, A. (1960) *Incorrigible Rebel*. London: MacGibbon & Kee.

Hunter, L.C., Reid, C.C. and Boddy, D. (1970) *Labour Problems of Technological Change*. London: Allen & Unwin.

Jacobi, O. (1983) *Economic Development and Trade Union Collective Bargaining Policy in the Federal Republic of Germany Since the Middle of the 1970's*. Frankfurt: Institut für Sozialforschung.

Kinnersly, P. (1974) *The Hazards of Work: How To Fight Them*. London: Pluto Press.

Markovits, A.S. (1986) *The Politics of the West German Trade Unions: Strategies of Class and Interest Representation in Growth and Crisis*. London: Cambridge University Press.

Miller, D. (1982) 'Social partnership and the determinants of workplace independence in West Germany', *British Journal of Industrial Relations*, 20: 44–66.

Prest, M. (1978) 'Union role may clinch Tenneco deal', *The Times*, 28 June.

Schuster, D. (1973) *The German Labour Movement*. Druckhaus: Deutz.

Streeck, W. (1981) 'Qualitative demands and the neo-corporatist manageability of industrial relations', *British Journal of Industrial Relations*, 19 (3): 135–69.

Streeck, W. (1985) *Industrial Relations and Industrial Change in the Motor Industry. An International View*. University of Warwick: SIBS&IRRU.

Thesen zur Gewerkschäftlichen Angestelltenarbeit (1986). Hanover: V.B. Finanzen-Angestellte. Nos 3–4.

Transport and General Workers Union typescript, distributed as 'Cyclo No: 841877' December 1984.

Wigham, E. (1968) 'Moving towards productivity bargains', *The Times*, 27 September.

9

Self-regulation under Pressure: Environmental Protection Policy in the Chemical Industry and the Response of the Sectoral Business Associations

William E. Paterson

Business interest associations in the European chemical industry have traditionally been well organized and enjoyed a harmonious relationship with governments of varied complexions. The centrality of the chemical industry, its technological achievements and good labour relations have combined to ensure that it has been largely insulated from political turbulence and unwelcome governmental intervention.

This very favourable balance for the industry has been disturbed by the increased salience of the environmental issue which seriously weakened the capacity of the industry to preserve its autonomy from governmental intervention. This chapter will focus primarily on German chemical business interest associations since they have been most affected by these developments, and their British equivalent, which has until now been under relatively little pressure.

For most of their history the associations devoted relatively little policy attention or personnel to the issue of environmental regulation. Their work was focused rather on trade policy, protection of patents and, especially in the German case, the forging and management of links between the chemical industry and the wider scientific community.

The primacy of self-regulation

The initial expansion of the chemical industry in the nineteenth century was associated with potential and actual risks to the environment both inside and outside the chemical plants themselves. States faced with expanding urban populations were initially concerned over the risks posed by discharges from chemical plants into the air and the aquatic environment. The most visible pollution was caused by the Leblanc process for the production of alkalis, which involved very large-scale emissions of hydrogen chloride into the air. This led in Britain to a series of Alkali Acts and the establishment of the Alkali Inspectorate in 1863; the first pollution control agency in the world. This early regulation by the state was in general not very onerous and the 'green-field' nature of the large chemical sites and the profitability of the industry meant it was able to cope with the demands imposed by the state relatively easily though Brock, the Managing Director of Sullivan and Co., claimed that compliance with the Alkali Acts had cost the industry £300,000 between 1863 and 1878 (Haber, 1988: 208). Attempts by environmentally conscious citizens who formed groups such as the 'Alkali Acts Extension Association' to tighten controls further were a failure.

The regulatory balance in any case began to shift from the late nineteenth century. In this period the chemical industry, especially the German chemical industry, underwent a sharp rise in public and elite esteem because of its spectacular scientific achievements; technological advances also helped to do away with the most visible emissions. The states themselves also became more and more dependent on the chemical industry. This was most obviously the case in Germany, where the dependence of the Imperial state on the chemical industry reached its zenith in the First World War and where IG Farben was central to the political economy of the Third Reich. This dependence by the state on the chemical industry was also evident in Britain, where a great deal of official effort was invested in trying to respond to the German lead in chemicals. The formation of ICI as a response to IG Farben had significant official support, and the leadership of ICI enjoyed a traditionally close relationship to Whitehall.

The undisputed technical brilliance of the chemical industry and the states' dependence on the chemical industry ensured that early measures such as the Alkali Acts did not develop into a tight regulatory regime for the chemicals sector. The pattern that developed was one of consultation between the state and industry with the overwhelming emphasis on self-regulation. This development was not, of course, confined to the chemicals sector, but the technical prestige enjoyed by the chemical industry resulted in a notably complete self-regulatory regime. In Dyson's terms it corresponded to the status preservation model which 'involves a politics of routine, day-to-day relationships, of quiet collaboration between "insider" groups and governments' (Dyson, 1982: 39).

This pattern persisted until relatively recently. In the Second World War the two states were again dependent on their respective chemical industries. The primary postwar imperative for both chemical industries was to expand production as part of, and a precondition of, the long-sustained postwar boom in the European economy.

The changing agenda

The Stockholm Conference on the environment, sponsored by the United Nations in 1972, failed to make the public impact its sponsors had hope for. However, involvement in the preparation of the conference had sensitized national bureaucracies to the increased importance of the issue, and the SPD/FDP coalition of 1969–82 in the Federal Republic took environmental concerns more seriously than the government of any other major state.

The new environmental policy of the SPD/FDP government was only occasioned to a minor degree by the external stimulus of the Stockholm Conference. The policy is much better understood as a response to changes in West German society in the late 1960s, especially the reordering of priorities among the younger, better-educated sections of society, away from the primacy of economics to quality of life issues such as participation and the environment. This value change reflected both the very high standard of living that the West Germans had by then achieved and the massive expansion of higher education.

In environmental policy, as in other policy areas, the two Brandt governments (1969–74) were in favour of more far-reaching reform than the three governments led by Helmut Schmidt. The Brandt government issued an Emergency Programme for Environmental Protection in 1970 and the First Environmental Programme in 1971. These initiatives stressed the 'polluter-pays' principle (*Verursacherprinzip*), modified, however, by the criterion of economic responsibility (*Kooperationsprinzip*).

These first laws were in general not sector-specific and did not pose a clear threat to the self-regulation model which remained unchallenged in the chemicals sector. The most important of these early laws were the Abfallbeseitigungsgesetz (waste disposal law) 1972, and the Bundesimissionsschutzgesetz (emission protection law) 1972. A much more threatening development as far as the industry was concerned was the move towards a more open, more anticipatory mode of decision-making in this policy sector. An Arbeitsgemeinschaft für Umweltfragen (working group for environmental questions) was established in 1971 to act as a source of policy advice for the federal government on environmental issues. The AGU incorporated all the relevant interests groups, including the, by then, rapidly expanding environmental groups. These groups were in a small minority in the AGU but they developed good relations with the Interior Ministry, the ministry with the greatest responsibility for environmental policy. To the alarm of the chemical industry it often found itself outnumbered by the representatives of the environmental groups when the Interior Ministry held consultative sessions on draft legislation proposals. In the VCI Report for 1974/5 the VCI complained bitterly that in relation to the proposed new TA Luft (technical instruction on air pollution control) industrial expertise had not been drawn on at an early stage (VCI, 1975: 47). Further nervousness on the part of the industry was occasioned by the foundation of a new institution, the Umweltbundesamt (Federal Environment Agency) in West Berlin in 1974. The establishment of such an agency appeared to be a signal that the government was no longer going to depend exclusively on 'the closed circle of experts' for advice on the formulation and implementation of environmental policy, but was going to go beyond industry and the established scientific community for advice.

These threatening clouds from the chemical industry's point of view were largely dissipated by the advent of the Schmidt government in 1974. The oil price rise of 1973 and the ensuing difficulties for the Western economies led to the reassertion of a relatively narrow view of economic priorities and a move away from an anticipatory, participatory mode of decision-making to a rather narrower focus on crisis management. The reassertion of economic priorities was a major theme at the so-called Gymnich discussions in 1975. Schmidt gave unequivocal priority to economic goals, and his views on environmental questions were much influenced by conversations with top union and management officials from Bayer. The pace of legislative enactments slackened, and the Abwasserabgabengesetz (waste-water tax law) 1976 and Bundesnaturschutzgesetz (Federal Nature Conservation Law) 1976 were notably weaker than their immediate predecessors. It also became increasingly clear to both the chemical industry and their environmental adversaries that the industry was in a position to ensure that its interests remained protected at the implementation stage, and disappointed environmentalists began to use the term *Vollzugsdefizit* (implementation shortfall).

In the SPD/FDP period there was one sector-specific law in relation to the chemicals area: the Chemicals Law of 1980. This Chemicals Law was the German implementing legislation of an EC Directive (Grant et al., 1988). Those political forces inside the Federal Republic, principally environmental groups – especially the BBU and the Interior Ministry – which would have liked a more stringent law, were easily overwhelmed by the Chancellor, the Economics Ministry and a parliamentary opinion which was predominantly sympathetic to the expert testimony of the chemical industry mediated through the VCI (Chemical Industry Association).

The break with self-regulation in the ensuing legislation was only minimal, however. The Chemicals Law is essentially a notification procedure where the firms carry out the great bulk of the tests on the new chemicals and the state regulatory agencies function as backstops. In this sense it is best seen as an example of 'regulated self-regulation'.

Self-regulation under pressure

The political climate which had favoured self-regulation for the chemical industry began to change quite profoundly after the election of March 1983. The changes were most profound at the level of the opposition, and I shall analyse these changes first.

The most significant change brought about by the election of March 1983 was the entry of the Greens into the Bundestag. The entry of the Greens had a dual significance. Firstly, it meant that there was now a party in the Bundestag which was unambiguously committed to an ecological perspective, and which would work consistently to push the regulation of environmental hazard higher up the political agenda. Its crucial significance was in the pressure that the Greens' electoral success put on the SPD's previous 'industry-driven' policy of the primacy of economics. A major aim of the SPD's electoral strategy was henceforth concerned with attempting to win Green voters back to the SPD. This concern pointed in the direction of more stringent environmental regulation of German industry.

There were now fewer obstacles to bringing about such a change. Helmut Schmidt had ceased to play any significant role in the SPD leadership after losing the Chancellorship as a result of losing a confidence vote on 1 October 1983. The veto on policy change in this sector imposed by the attitude of the unions was also weakening. IG Metall, the largest union, was now much less opposed to environmental measures. The new disunity left some room for manoeuvre for SPD policy-makers.

The loss of governmental power in October 1982 and the electoral defeat of the SPD led to a reorganization of the SPD Bundestagsfraktion. Two new fields were established, each headed by a Vice-chairman of the Fraktion in order to enable the SPD to offer more effective electoral competition to the Greens. These were Environmental Politics, headed by Volker Hauff, and Women's Rights and Equality, led by Helga Däubler-Gmelin.

The Working Group on Environmental Politics quickly became interested in environmental control of the chemical industry. The antinuclear theme at that time was occupied fairly comprehensively by the Greens and it took Chernobyl to dislodge the SPD finally from its pronuclear stance. In contrast, the field of chemicals control appeared to offer an electorally promising alternative where the Greens had been relatively inactive. Environmental groups had been quite active in this area. Bund (the League for Environmental and Nature Protection in Germany) had first carried the term Chemiepolitik (chemicals policy) in 1983 and published their first position paper, *Chemiepolitik – Bund fördert einen neuen Politikbereich* (Chemicals Policy – Bund requests a new policy area) in 1984. The Working Group on Environmental Politics rapidly established a subcommittee on Health and the Environment which was co-chaired by Michael Müller, a former official of IG Chemie, and Freimut Duve, a leading figure on the SPD's environmentalist wing. This subcommittee then established a further subcommittee on chemicals and the environment.

In the 10th legislative session 1983–87 the SPD produced a number of initiatives in relation to chemicals control at the federal level. The most important initiative was the so-called chemicals policy (Konzept für eine Umwelt and Gesundheitsverträgliche Chemiepolitik) which was published by the SPD in April 1986. This policy was basically concerned to tighten up the procedures for the introduction of new chemical substances. Its major innovation was the concept of social net benefit (Sozialer Nettonutzen). This concept specifies that new products should be introduced only if their production represents the minimal possible danger to plant workers and the environment, if their introduction would lead to lesser use of resources in production, and if their production presents the least possible danger to health and the environment. The chemicals policy also included proposals for the identification and listing of the most suspect of existing chemicals and the conclusion of a voluntary agreement with the chemical industry on the notification of intermediate products.

The suggested policy also contains very detailed suggestions on ways of improving the rights of employees in relation to the handling of dangerous chemicals. These proposals were clearly designed to bind IG Chemie to the chemicals policy.

Out of power in Bonn the SPD were still in power in five Land governments. This gave the SPD an alternative platform for environmental initiatives. The major thrust of the SPD at the Land level was a joint initiative for a tax to be paid by the chemical industry to pay for the clearing up of chemical dumps. The SPD, although in a stronger position at the Land level than in the Bundestag, were still in a minority in the Bundestag and there was little chance of this initiative getting on the statute book.

The Greens

The entry of the Greens into the Bundestag in 1983 posed significant potential problems for the chemical industry. In practice, however, the Greens concentrated on nuclear energy and peace issues in the Bundestag and the major conflict between the Greens and the chemical industry occurred at the Land level when a leading Green, Joschka Fischer, became Environmental Minister in Hesse in December 1985. Fischer concentrated on much stricter implementation rather than new legislation, and was able to exert considerable pressure on Hoechst during his incumbency of just over a year. His decision to concentrate on the implementation stage, his disinclination to proceed in a 'cooperative' manner and his readiness to hire new civil servants less sympathetic to industrial needs all seemed ominous to the VCI.

The response of the VCI

Confronted by an increasingly adverse domestic climate the West German chemical industry has maintained its commitment to self-regulation as its strongly expressed preference for dealing with environmental hazard. Its second preference is for voluntary agreements and a large number of voluntary agreements have been concluded especially, but by no means exclusively, in the area of plant safety; for example reduction in the use of CFCs and NTAs in detergents. Its third preference would be for EC- or OECD-wide measures of environmental regulation, and its least preferred solution is for legislation restricted solely to the Federal Republic.

In defending its core position the VCI normally deploys three kinds of arguments. The first set of arguments stresses the damage to the competitiveness of the German industry posed by heavy environmental costs. This argument is sometimes flanked by an implied threat that the industry will choose to expand elsewhere. The interlocking and high-cost nature of chemical plants weakens the force of this auxiliary argument. It has, however, featured prominently in the announcements by both Bayer Chemicals and BASF that they intend to concentrate the fast-expanding area of gene technology in the United States rather than the Federal Republic. A second set of arguments stress what the industry is already doing. In 1986 it claimed to be spending DM4.5 billion on environmental protection. These arguments also point to what the industry considers successful examples of self-regulation, for example the TUIS (Transport Accident Information Scheme) on the transport of hazardous substances. The final set of arguments stress the threat to innovation imposed by the need to conduct a high proportion of R&D in a defensive mode to conform with bureaucratically imposed tests and regulations. This argument is often made with reference to specialty chemicals which are often produced in small quantities. These arguments represent the VCI's core position but are presented differently depending on the target audience.

The priorities of the CIA differ significantly from those of the VCI. It has the same commitment to self-regulation and its second preference would be for voluntary agreements. It is much more unhappy about EC-wide measures of environmental regulation than the VCI and, unlike the VCI, its very strongly expressed preference would be for legislation confined to the United Kingdom. In defending its position the CIA makes use of very similar arguments to the VCI with an even stronger emphasis on the dangers to the competitiveness of the chemical industry since the British industry, unlike the German, is not heavily involved in the production of clean technology.

The VCI has made a very considerable investment since 1979 to influence public opinion. 'All the Greens' news automatically gets into the newspapers. I have to pay for advertisements' (Wolfgang Munde, VCI). The advertisements are financed by the Initiative Geschützter Leben. In attempting to reach public opinion the VCI has not relied on its total repertoire of arguments but has attempted to stress the positive achievements of the chemical industry in improving the environment. This approach has been very ineffective, and one television picture of a chemical spillage undoes months of investment in expensive advertisements. Since 1988 the large companies have made a much more regular and intensive practice of producing large-scale advertisements setting out what they have done individually for environmental protection.

In dealing with governmental bureaucracy the VCI normally stresses the first and third sets of arguments, that is the dangers to competitiveness and innovation.

The style of argument deployed by the CIA is much more ad hoc and less easy to characterize. The CIA has been very concerned to monitor public opinion in relation to the chemical industry, and it certainly perceives the findings of successive surveys as worrying. Despite these worries, it has not at the time of writing established an equivalent to the Initiative Geschützter Leben, though plans exist to sponsor a series of advertisements. The generally cooperative nature of the relationship between the British chemical industry and the state means that it has not needed to develop a highly articulated set of arguments against environmental regulation. The arguments it has developed are counter-arguments to defend the specific interests of the British chemical industry against what is often seen as the too-ambitious EC proposals on the environment.

The organizational reaction of the business interest associations

The response of the business interest associations to these changes has diverged considerably. Organizationally the VCI has undertaken a much longer commitment than the CIA, reflecting both the much greater saliency of the issue in the Federal Republic and the greater scale of resources available to the VCI. The major change was to create a new department of technology and the environment to work under the direction of a new similarly named committee, which superseded the long-established Technology Committee. This department is now the largest in the VCI, servicing eight committees, fifty-three working groups and eighteen technical working groups on individual substances. The VCI now claims that one-quarter of its personnel work in one way or another in connection with the environmental issue.

The VCI also established a special organization, the Initiative Geschützter Leben, in 1979 to attempt to influence public opinion. This initiative is financially supported on a voluntary basis by sixty firms and eleven Verbände. Its chairman is usually the Vice-president, that is the immediate past President, of the VCI.

Changes within the Chemical Industries Association have been more modest, reflecting both the smaller size of the CIA and the lower political saliency of environmental issues in the United Kingdom. The CIA appointed a Director for Health, Safety and Environment. The principal CIA committee in this area is the CIHSEC (Chemical Industry Health Safety and Environment Council) which works alongside the Trade Affairs Board and the Industrial Relations Board in the CIA structure. The CIA has only one full-time official working on environment questions narrowly defined, though there are a total of five officials in the CIA Health, Safety and Environment Directorate.

Specialist associations

Environmental issues vary considerably in importance to the various subsectors within the broad chemical sector. It is an especially important issue for agrochemicals and the detergents and toilet preparations sectors. This means that the specialist associations in these areas cannot simply rely on the CIA and VCI to defend their specific interests as part of a general defence strategy, but have to take their own specific initiatives in relation to issuing material in defence of contested issues, such as animal testing for the toilet preparation industry and lobbying. The German specialist associations still rely very heavily on the VCI lobbyist as they have no Bonn presence, but the importance they attach to defending their position on environmental matters can be gauged from the fact that the Industrieverband Pflanzenschutz (German Agrochemical Associations) pays for 30 per cent of the total time of the Bonn lobbyist of the VCI, and uses this time very largely for lobbying on environmental issues.

Peak associations

Whilst the sectoral business associations share the task of defending the industry on environmental policy with selected specialist associations, they also rely on the peak associations, the BDI and the CBI to defend their position. They try to ensure that the positions are coordinated and a representative of the chemical industry is chairman of the BDI's Environmental Committee. A relatively large proportion of environmental legislation is non-sector-specific, and it is important for the chemical industry to make its voice felt as part of a wider industry response.

The search for allies

Both the CIA and the VCI look primarily to the governmental level for allies in resisting far-reaching environmental legislation. The CIA focus is on the Department of Trade and Industry which acts as a sponsoring department for it. The concept of sponsorship would certainly include defending the industry against too stringent environmental regulations.

Broadly speaking a similar picture emerges in the Federal Republic and the VCI's main support within the government is the

Ministry of Economics, especially the chemicals section which acts as a clearing-house for government policy affecting the chemical industry. The German chemicals industry now also receives a great deal of support from the Research Ministry which is headed by a chemist, Dr Riesenhuber, and which takes an explicitly industry-friendly line. In both countries the importance of the agro-chemicals sector has led to the development of close relationships between the chemical industry and the respective agricultural ministries.

The major difference in the orientation of business interest associations in the chemical sector towards government in the two systems derives from the federal nature of the West German system. The VCI has a territorial dimension of regional associations and they are sometimes used by the VCI as a secondary route to enlist the support of Länder governments, for example in the transformation of the EC Sixth Amendment into the German Chemicals Law in 1979/80 the VCI made a large number of representations to Länder governments which did lead to significant amendments being made when the legislation was considered in the Bundesrat. The support of Länder governments is very important, since nearly all legislation now requires the consent of the Länder expressed in the Bundesrat. Legislation is also considered first in the Bundesrat and then in the Bundestag. Nearly all Bonn ministries possess little or no field service, and legislation is implemented by Land bureaucracies. This gives the VCI a vital interest in terms of the specialist enabling legislation which is produced for each Land, and in maintaining contact with the Land bureaucracies. This is especially so in a regulatory field like environmental policy, where the attitudes of public officials at the implementation stage make a vital difference to the impact of public policy.

There are more pronounced differences in the way in which the associations handle the search for allies in their respective legislatures. These differences have become more marked over recent years. Until relatively recently the legislature was very much a secondary target for the associations in comparison to the bureaucracy. The pharmaceutical subsector is a partial exception, and has had to devote more resources to cultivating relations with the legislature than the remainder of the chemical sector. This is still broadly true in Britain, where environmental policy has been much less politicized than in the Federal Republic.

In West Germany the Bundestag occupied much the same marginal position as Parliament in the United Kingdom. The key interlocutor was the bureaucracy, since legislative proposals in the

Federal Republic normally emerge as a product of the interaction between interest groups and the bureaucracy. Moreover, it is bureaucrats rather than legislators who are the key actors at the vital implementation stage. The perception of the Bundestag as marginal is now changing, as the VCI becomes more aware that the high political visibility of environmental policy is pushing it much further into the directly political arena and away from its former neutral/bureaucratic arena. This gives the views of the political parties a much heightened importance, and means that the VCI has to spend much more time in cultivating them and parliamentary opinion.

In cultivating the political parties a second difference emerges. The CIA has generally pursued a non-partisan strategy and has sought support from all the parliamentary parties. This was for a long period also the strategy of the VCI, reflecting the basic consensus on the economy that obtained between the established German political parties until the 1980s. Since then the VCI has adopted an increasingly partisan stance. The FDP and CDU/CSU are perceived as unequivocal allies, with the FDP being especially valued as an ally both because the Economics Minister has been from the FDP since 1974 and because of the closeness of the party's market ideology to the core values of the chemical industry. The partial conversion of the SPD to ecological positions has resulted in the SPD no longer being regarded as a reliable ally and the VCI, as in the 1987 election, publicly endorses the CDU/CSU and the FDP. In its dealings with the SPD the accent is now on damage limitation, on containing the environmental drift in the SPD's policies. In pursuing this policy the VCI is able to count on the support of IG Chemie, whose chairman, Hermann Rappe, is an influential member of the SPD's Bundestagsfraktion. The VCI has also been helped by the fact that the Bundestagsfraktion has been much more 'industry-friendly' than the party as a whole. This is starting to change in the wake of the 1987 federal election, when 25 per cent of the new Bundestagsfraktion are in the Bundestag for the first time.

The VCI still invests resources in cultivating the SPD and trying to dissuade it from what the industry sees as foolish and ill-considered policies. If the SPD, either at federal or Land level, develops an initiative which will affect the chemical industry, the VCI is always ready to respond to SPD requests for meetings and invariably sends a mixed team of the most important VCI office-holders, company experts and VCI officials to the meetings.

With the Greens, in contrast, the relationship is one of unrelieved hostility. No effort is spent in cultivating the Greens and no delegations are sent to meetings with the Greens, lest it lend further weight to the Greens' position.

The relationship with the labour unions

Arguably the greatest difference between the CIA and the VCI/BAVC in relation to handling the issues raised by demands for more stringent environmental regulation lies less in the way they handle relations with government and more in their relations with the trade unions.

In Britain the general relationship between capital and labour in the chemical sector has been good, and industrial conflict has been fairly low in comparison with other sectors of British industry. The field of health, safety and environment is something of an exception, however, and relations with the labour unions in this area have been relatively adversarial.

The leadership of the industry and the CIA have not actively sought to enrol the unions in defence of the industry against more, or, as they would see it, too stringent environmental regulation. The concern of the CIA has been rather to limit the involvement of the unions. They adduce three reasons for this strategy. The absence of an industry union on the German pattern means that the amount of relevant expertise is spread very thinly among the unions inactive in the chemical sector. The first argument of the industry is that the unions are simply not competent to judge the technical issues involved. CIA representatives argue secondly that the unions' interest in environment is purely tactical and that they never pursue initiatives with any vigour. It is mainly a tactic in this view to extract concessions elsewhere. The final argument would be that pressures for stricter environmental regulation in Britain are not internally generated but trickle down from the EC. In this situation the trade unions are not important potential allies, and efforts should be concentrated on the governmental level in the UK and in CEFIC.

The chemical industry has pursued a strategy of minimal involvement of the trade unions very consistently. The Health and Safety Commission and its executive arm, the Health and Safety Executive, are tripartite bodies, but there has been a long-running controversy about the failure of the Health and Safety Commission to establish an Industrial Advisory Committee for chemicals. This has not been settled by setting up a National Joint Advisory Committee on Health and Safety in the Chemical Industry. This committee includes the national officers of the relevant unions and senior management figures, but most union opinion considers it to be less impressive than analogous bodies for other industries. It is relatively rare for unions to be consulted at anything other than a very late stage in this area.

The unions have reacted to this exclusion by continually articulating their worries about the standards of environmental protection obtaining in this area. Their primary accent has been on the effects on their members, but they have also shown a concern for the wider effects; for example in December 1986 the Transport and General, the General and Municipal and NUPE launched a joint campaign on the use of pesticides. This campaign concentrated both on the effects on their members (primarily farmworkers) and the public at large (*Sunday Telegraph*, 7 December 1986). The effect of these campaigns is difficult to gauge. The unions are widely distrusted and the campaigns are both poorly funded, and pursued without great energy. They have also had a very limited impact on the wider labour movement which is much less environmentally conscious than its German counterpart.

As in Britain, the general relationship between capital and labour in the German chemical industry is markedly harmonious, even in the context of the generally highly consensual character of industrial relations in West Germany. Unlike Britain, this harmony also prevails in relation to environmental policy. The overriding priority for the VCI/BAVC and IG Chemie is defence of the industry, and it would not be an exaggeration to speak of a 'producers' alliance' in this area. In broad areas of policy for the chemical industry IG Chemie is hardly consulted, but in the area of environment and health and safety quite a high degree of coordination and cooperation takes place between employers and unions to defend the industry against what they both perceive as unreasonable pressure from environmentalists. This coordination takes place in the form of continual consultation, regular meetings at the most senior level with both VCI and BAVC and with the Bonn lobbyist of the VCI.

For a considerable period IG Chemie statements on environmental policy were more or less identical to those of the VCI, and deployed the same core arguments to defend the industry. In addition they more consistently referred to the importance of the safety at work dimension of environmental policy. This emphasis posed few difficulties for the employers, since the German chemical industry has, and can afford, very high occupational safety standards. IG Chemie also used the theme of safety at work to attempt to extend codetermination rights, a demand which has met a great deal of resistance from the employers.

By the early 1980s the environmental cleavage line also began to run through the DGB and the SPD itself. Inside the DGB a number of unions, especially IG Metall, attempted to absorb some ecological positions. The defeat of Helmut Schmidt in October 1982, and the Greens' victory in March 1983, led to a widening of SPD perspectives which included more environmentally friendly positions.

IG Chemie, in order to preserve its influence in what now became for the VCI the opposing camp, had to modify its position. It continued to defend the chemical industry but could not be seen any longer to oppose all moves for the stricter environmental regulation of the chemical industry. Its major thrust is against any accommodation between the SPD and the Greens, and it uses every possible public statement to rule out such an accommodation.

In the public discussion on the Rhine spillages of Autumn 1986 IG Chemie expressed no public criticism of the industry; indeed the union had a conference at the height of the spillages emergency attended by major figures from the industry where the dominant impression was one of solidarity.

IG Chemie responded warmly to a suggestion by the BAVC that the Umweltbeauftragter (official responsible for the environmental questions) in each company be placed on the board. They also proposed establishing an Environmental Foundation together with the employers. Determined attempts to extract concession from the employers on codetermination by extending the functions of the Economics Committee (equal representation of capital and labour) to include environmental matters and strengthen the rights to information have not so far been successful (*Handelsblatt*, 4 May 1987). The foundation GIBUCCI was established, however, in January 1988. It is jointly sponsored by the industry and the union, and its role is to inform workers' representatives in the plants about environmental protection.

The collapse of the self-regulation strategy in the FRG

The VCI responded to the increasing pressures from its domestic critics by developing a set of Leitlinien (guidelines on environmental policy) which it issued in July 1986. The guidelines were greeted with some enthusiasm by Walter Wallmann, the newly appointed Environmental Minister. The strategy was completely undermined by the series of chemical spillages into the Rhine in November/December 1986. Not only were all three major German chemical companies involved, but it occurred at the peak of the federal election campaign. This combination seriously weakened the political allies of the industry, including Walter Wallmann, and severely compromised the chances of maintaining a policy based on self-regulation.

Business interest representation at the European level

The chemical interest associations have had to respond in the last decade and a half to an accelerating shift away from purely national policy-making in relation to environmental policy. The shift has been driven by the converging interests of three different actors, the EC Commission, the chemical industry and the Federal Republic of Germany.

The establishment of the Common Agricultural Policy and the customs union in the 1960s and the Gaullist check on its political ambitions, left the EC Commission looking for new policy areas. This impulse coincided with a general increase in interest in the environment, dramatized but not created by the Stockholm Conference. After a protracted series of negotiations between the Commission and the member states agreement was reached in March 1973 that the Commission should receive prior notification before the introduction of any significant new policy proposals on the member states to protect the environment. The Commission can block the development of a unilateral national initiative by developing a proposal of its own within five months. In 1978 the Commission's capacity to develop policy in this area was significantly increased by the establishment of a new directorate (DG11) to handle environmental policy.

The second pressure away from national solutions in relation to environmental control in the chemicals sector derives from the international character of the industry. This gives the industry an interest in a common set of conditions for registration of products in order to avoid the costly duplication of tests. It has a further interest in developing a common EC position in order to negotiate a common harmonized position for Western Europe and the United States. The OECD provides a convenient forum for such negotiations.

The last major actor pushing for European-wide regulation in this area has been the government of the FRG. The first preference of German industry in general and the chemical industry in particular had been self-regulation. This option came under increasing pressure in the late 1970s. Externally, the adoption of TSCA in 1976 by the United States made the development of regulations to cover market entry for new chemical substances a priority in order to be able to exert some leverage in the proposed American legislation by negotiation. Internally, the strength of environmentalist pressure was making the maintenance of self-regulation increasingly difficult. This clearly constituted a threat to the competitiveness of German industry, and the initial attempts by the federal government to create European-wide regulatory regimes were clearly a response to this. In recent years West German industry has invested a great deal in clean technology and would stand to gain from the EC-wide adoption of high environmental standards. In the chemicals sector, Degussa has a commanding position in the production of catalysts to reduce vehicle emissions and Uhde, a Hoechst subsidiary, is a leading manufacturer of pollution control equipment.

The increasing concentration on the EC level as the locus of policy-making on environmental policy led to a dual response on the part of the business interest associations. The Council of Ministers of the EC continued, despite the intentions of the framers of the Rome Treaty, to operate on the basis of unanimity. This allowed individual member states considerable veto powers and reinforced the continued importance of relations with national governments for the business associations. In handling the increasing proportion of environmental issues possessing an EC dimension, business interest associations pursue two routes.

The national route

The increasing importance of the European level has not rendered the national level unimportant, and business interest associations continue to attach enormous importance to the national level. This is obviously especially true when the views of different national chemicals industries diverge. This is very obviously the case in relation to a number of environmentally related issues. The British chemical industry, faced with a much easier domestic political response on this issue and with regulatory assumptions which less clearly fit EC models, is much more hostile to environmental legislation than its German counterpart.

> Without an effective enforcement capability of its own, the Community has evolved a legalistic approach to harmonization in all areas, spelling out in precise rules the criteria for compliance by its member governments and constituents. In toxic substances policy, this approach favours such regulatory techniques as numerical tolerance levels, lists of approved substances, and detailed guidance on such matters as packaging and labelling. A more informal, voluntary approach to toxic substances policy, such as that followed in Britain, is scarcely suitable as a model for international policy harmonization. (Brickman et al., 1986: 290)

The British chemical industry now often find themselves isolated on environmental issues, with little support either from DG11 of the Commission or from the other national chemical industries. In this situation the CIA relies on its close association with the DTI to ensure that its viewpoint is articulated by the British government in EC negotiations. The German industry still clings to self-regulation as its preferred option, but, over the past decade and a half, this has, given the high environmental consciousness of values in the Federal Republic, become increasingly difficult to sustain. Its reluctant preference is therefore for action at the European rather than the national level. In attempting to influence outcomes at the European level the VCI is greatly aided by the fact that its closest political ally, the Economics Ministry, is also the ministry charged with the coordination of day-to-day European policy.

The European route

The key role of the Commission in the formulation and preparation of EC directives, and the need to coordinate and, if possible, aggregate the differing positions of national chemical industries, led to the establishment of CEFIC, the European Council of Chemical Manufacturers Associations. Environmental issues in CEFIC are the responsibility of the Technical Affairs Department. The work of the Department is organized through five principal committees, the Health Protection Committee, the Safety of Production Committee, the Safety of Chemicals Committee, the Transport Committee and the Environment Protection Committee.

CEFIC invests heavily in time and resources in attempting to influence the Commission on environmental policy. It has especially intensive relations with DG5 (Employment and Social Affairs), DG6 (Agriculture) and DG11 (Environmental Protection).

CEFIC has hitherto been very well placed to exert influence on the Commission. The Commission has itself been in a weaker situation than in some other policy sectors because of the absence of specific references in any of the founding treaties to environmental policy. This constrains the Commission to act cautiously and to rely on general strategies such as harmonization.

The Single European Act now provides specifically in Article 130 for environmental policy. This article, moreover, explicitly endorses the 'polluter pays' principle. The explicit endorsement of environmental policy has engendered fears that the environmental policy sector will become much more politicized, and that the Commission will make much greater efforts to involve environmentalist groups in its policy network. One possible route for this is through the European Parliament. Fears that this might be the case are reflected in the final form of Art. 130(S) where the European Parliament's rights of consultation are weaker than in many other articles of the Treaty. Nevertheless, CEFIC is notable among European-level interests groups for the time and resources it devotes to liaising with, and monitoring the activities of, the European Parliament.

The Health Protection Committee is an advisory committee of chief medical officers and industrial hygienists which meets infrequently. The other committees involve a high degree of participation by company experts. The Safety of Production Committee is concerned with health and safety at work, with a major focus on ten implementations of the Seveso directive. The Transport Committee coordinates activity in the areas of transport and distribution. The Environment Protection Committee is concerned with the emission of toxic waste and environmental policies. The Safety of Chemicals Committee concentrates on classification (carcinogens), existing chemicals, directive preparations, confidentiality, animal welfare and export ratification. In practice much of the detailed work is done by specialized working parties and task forces involving experts from the companies and national associations.

Hitherto industry has enjoyed its clearest advantage over its environmental adversaries at the implementation stage where environmental groups have been unrepresented. CEFIC is therefore especially concerned about changes in the Commission's attitude towards implementation revealed in its Fourth Environmental Action Programme (1987–92). Hitherto the Commission has regarded implementation almost exclusively from the perspective of legal compliance, that is whether the state authorities have enacted satisfactory domestic legislation or not. This approach had two consequences. Firstly, the Commission often became engaged in protracted negotiations with states that had enacted legislation that was not totally satisfactory, but ignored countries which had introduced no legislation at all. This problem continues to increase as a result of the southern extension of the EC. Secondly, there is a danger from the point of view of the chemical industry that complaints about compliance will mirror the respective strength of the environmental movements in the various members states rather than the actual degree of compliance.

Conclusion

The main argument of this chapter has been that the self-regulation model, which has been the centre-point of the chemical industry view of government/industry relations, has been severely tested by the increased political saliency of the environmental issue. This is most clearly the case in the Federal Republic but it is also increasingly the case in relation to the British chemical industry. The British strategy of resisting EC environmental issues looks increasingly unpromising given the Single European Act and a number of issues including fluorocarbons. Britain has recently had to accede to EC agreements after years of resistance ('Dirty old Britain cleans up its act', *Financial Times*, 21 March 1987).

Increased public intervention and regulation will not be constant across the whole area of chemicals production. It is likely to continue to be at its most intense on issues such as emissions into the air and the aquatic environment. On the products themselves it is difficult to conceive such a dramatic shift, since the regulatory authorities are necessarily and inevitably (without an unimaginable increase in their staff) almost totally dependent on the expertise of the producers.

Even in the areas which have had the highest political saliency such as emissions, and where the industry has been subject to external regulation, the domination of the implementation stage by experts from the governmental bureaucracy and the firms has in practice often come close to self-regulation. The industry is therefore especially concerned by the Fischer example, where the emphasis was precisely on the implementation stage and it is even more concerned by the proposals in the Fourth EC Environmental Programme to strengthen the monitoring of implementation. The more salient the environmental issue becomes, the more difficult the sectoral associations will find it to preserve a depoliticized climate, and the more likely they will face regulators who are determined to impose existing regulations upon them.

Perhaps the most interesting question is the extent to which the partial collapse of the self-regulation model in the field of environmental regulation will lead to a wider redefinition of the relationship between sectoral business interest associations and government. My impression from the British and German experience is that this would not be the case. The political complexion of the British and German governments makes them very unlikely to utilize the weakness of the industry in this area as a point of entry for a redefinition of the wider government–industry relationship.

References

Brickman, R., Jasanoff, S. and Ilgen, T. (1986) *Controlling Chemicals: The Politics of Regulation in Europe and the United States*. Ithaca: Cornell.

Dyson, K. (1982) 'West Germany: the search for a rationalist consensus', in J. Richardson (ed.), *Policy Styles in Western Europe*. London: Allen & Unwin. pp. 32–64.

Grant, W., Paterson, W. and Whitston, C. (1988) *Government and the Chemical Industry*. Oxford: Oxford University Press.

Haber, L. (1988) *The Chemical Industry During the Nineteenth Century: A Study of the Economic Aspect of Applied Chemistry in Europe and North America*. Oxford: Clarendon Press.

VCI (1975) *Jahresbericht des Verband der Chemischen Industrie*. Frankfurt: Verband der Chemischen Industrie.

10

The Overcapacity Crisis in the West European Petrochemicals Industry

Wyn Grant

This chapter is concerned with examining the serious overcapacity problem which afflicted the West European petrochemicals industry in the early 1980s, and the mechanisms through which it was largely resolved. After a period of prolonged and rapid growth the industry was faced with a decline in the rate of growth in demand for its products (or even an actual fall). At the same time as the position on the demand side was deteriorating, new sources of supply utilizing cheap energy feedstocks were coming into operation outside the traditional production areas of Western Europe, the United States, Japan and the state trading countries. Particular attention focused on Saudi Arabia but, as will be discussed in more detail later, there are many other new producers coming on to the market. The long-run impact of these new producers is not likely to be felt so much in terms of imports into Europe, although the imports which have taken place have attracted considerable attention, but the displacement of European exports from third country markets. The consequence will be a substantial erosion of the West European balance of payments surplus in petrochemicals.

This analysis will not be concerned so much with the substance of the problem, as with the mechanisms through which attempts were made to resolve it. In particular, attention will be focused on the intra-European aspects of the problem. Of course, the issue obviously involves questions which concern Western Europe's trading relations with other trading blocs. In the past, for example, much attention has been focused on the impact of increased chemicals production in Eastern Europe and the Soviet Union; in the 1980s the petrochemicals question became a key element in the Euro-Arab dialogue. With the changes in Eastern Europe, the impact of production there reappeared on the agenda in the 1990s. These wider ramifications of the problem will be mentioned, but they will not be the central focus of the discussion. In line with the general concerns of this book the principal theme will be the extent to which associative action can help to resolve the problem of overcapacity. The discussion will cover both the extent to which it has made a contribution, and whether associative action has an unrealized potential for problem-solving, bearing in mind Kenis's (1985) argument that it was underutilized in tackling the overcapacity problem in the chemical fibres industry.

The nature of the crisis

As will be discussed in more detail later in this chapter, an overcapacity crisis in the European petrochemicals industry which had appeared to be extremely serious at the beginning of the 1980s had at least been largely resolved by 1987. As one commentator noted: 'the industry has very nearly succeeded in achieving balance between supply and demand in many important product areas' (Wamsley, 1987a: 3). Admittedly, there were still grounds for concern about longer-run trends. It was also the case that 'The necessary reductions in capacity were not achieved without a great deal of pain' (Wamsley, 1987b: 3). Even so, the position of the Western European petrochemicals industry compares favourably with that of the steel industry:

> By the winter of 1983–4 . . . the European steel industry was in a disastrous state, with no country and no firm (private or public) spared, with punishing losses and colossal debts, with virtually new plant being scrapped or moth-balled and all plant being under-utilised, and with steel communities left devastated. Moreover, in spite of huge capital write-offs and a vast and costly adjustment programme involving modernisation, rationalisation and savage contraction the prospects for the decimated industry and workforce remained gloomy (Mény and Wright, 1986: 10)

Before outlining the various mechanisms which were available, one general point should be made. The problem was a West European one which had to be resolved at a West European level. ('West European' in this context does not just mean the EC, as is evidenced by the participation of non-EC countries in the European chemical industry organization, CEFIC. However, in practice, the major EC chemicals producers and Community institutions take a leading role in dealing with problems of this kind.) Indeed, in many respects it was a global problem, but the differences of interest between American and European producers made any effective concerted approach a non-starter. The problem had to be tackled at a European level because the petrochemicals industry is a highly internationalized one in which Western Europe is regarded by the major companies as one market. It is not a perfectly integrated market; there are national divergences. As the former Chairman of ICI has commented:

> Europe is still a collection of national markets to a very considerable degree. . . . We still find it necessary to serve individual countries with individual products although that, hopefully, is breaking down. (House of Lords, 1985: Q1346)

There were six main options available to the industry in dealing with its problems in the 1980s:

1 Unilateral actions by individual companies to close plants.
2 Joint actions by companies involving portfolio exchanges or joint ventures. Joint ventures were a familiar form of cooperative action in the European petrochemicals industry, but portfolio exchanges were a new idea.
3 Cartelization.
4 Actions by national governments.
5 Actions by the EC.
6 Facilitative action by associations.

Each of these options will be dealt with in turn in some detail. However, the overall position will be briefly outlined here. First, a number of companies did take unilateral actions to close plants; however, by themselves, these would not have dealt with the scope of the problem. Second, joint actions by companies made the largest single contribution to bringing the problem under control. The third option, cartelization, is both illegal and less feasible than it was in the inter-war period when there were fewer players in the market; that is not to say that it was entirely absent, as EC action against a polypropylene cartel shows. National governments did become involved in the problem, if only because of the importance of state holding companies in the industry, but, because of the international character of the industry, their scope for effective action was limited. As far as the EC was concerned, the idea of a 'crisis cartel' was ruled out at an early stage, but the EC did facilitate joint action by companies to cope with the problem. Facilitative action by associations did help to contribute to a climate of awareness of the seriousness of the problem, and particular attention will be paid in the ensuing discussion to the role of the new Association of Petrochemicals Producers in Europe (APPE).

Before discussing each of these options in detail it is necessary to explore the nature of the overcapacity problem in greater depth. There is no doubt that at the beginning of the 1980s the West European petrochemicals industry faced a grave crisis. Following

Table 10.1 *European capacity utilization (nameplate capacity) (percentages)*

	1981	1985
Ethylene	71	84
LDPE/LLDPE	62	89
HDPE	64	84
PP	75	96
PVC	64	82
PS	60	83

Source: Greenwell Montagu Research.

approximate growth rates of 25 per cent per annum in the 1950s and 15 per cent per annum in the 1970s, demand for the bulk polymers in Western Europe fell by 10–20 per cent in 1981 (Beynon, 1982). The extent of the overcapacity problem at the beginning of the 1980s, and its subsequent easing, can be discerned from Table 10.1. Table 10.2 shows how, after a catastrophic fall in demand for bulk polymers in Europe in 1979–81, there was a compensating recovery in 1981–3, followed by very slow growth in 1983–4, and a considerable recovery in 1984–6. Even so, the negligible rate of growth in consumption between 1979 and 1984 was very difficult psychologically for an industry which had been used to operating in conditions of rapid expansion.

It is clear that there are variations in the positions of the different bulk polymers (ethylene is the basic 'building block' of the industry). Polypropylene has the best consistent performance, never falling below 75 per cent of capacity utilization and showing a 7.5 per cent average growth per annum in consumption between 1979 and 1984. It combines good impact strength, high softening point and low cost per unit of stiffness. It has widely based applications and is still finding new applications at the expense of other materials and plastics; it is, for example, extensively used in the auto industry. The greatest problems have been experienced with PVC, which is particularly used in the construction industry, and has been affected by the decline in demand for that sector. However, it was even possible to be relatively optimistic about PVC by 1987: 'there is now increasing evidence to suggest that even in this historically disastrous product area, supply/demand balance is close' (Wamsley, 1987b: 5).

Despite these divergences in the performance of individual polymers, the capacity utilization figures demonstrate the extent of the crisis facing the industry at the beginning of the 1980s.

Table 10.2 *Demand trends for commodity petrochemicals in Western Europe*

(a) Development of consumption in Western Europe, 1979–84 (change compared to previous year in percentages)

	1981/1979	1983/1981	1984/1983	Average growth 1979–84 p.a.
LDPE	−16	+10	−2.5	−2
HDPE	−10	+17	+6	+2
PP	+9.5	+20.5	+11	+7.5
PVC	−13	+10	+1	−1
PS[a]	−15.5	+13	+3.5	< +0.5
'The big five'	−12	+12.5	+2	< +0.5

[a] Excluding copolymers.

Source: Greenwell Montagu Research.

(b) Development of estimated total market in Western Europe, 1984–6 (change 1986 compared to 1984 in percentage)

	1986/1984
LDPE	+12.6
HDPE	+23.5
PP	+27.0
PS	+12.0
PVC	+12.0

Source: Calculated from Association of Plastics Manufacturers in Europe statistics.

Tables 10.2(a) and (b) are not strictly comparable, as they rely on different sources, but they provide a broad overall picture of demand trends.

Moreover, it should be noted that leading companies were making heavy losses on the production of petrochemicals; even in 1985, ICI were believed to be losing a million pounds a week on PVC. The factors contributing to this situation may be analysed from a demand-side and supply-side perspective.

From a demand-side perspective the petrochemicals industry faced a number of problems at the beginning of the 1980s. First, average growth rates in Western Europe were below those experienced before the first oil shock. As an 'industry's industry', the chemical industry is substantially affected by the rate of growth in the economy as a whole. Second, the industry was facing increasing technological maturity. Most of the significant innovations in petrochemicals were made before 1960 on two fronts: new products (polypropylene was the last major bulk polymer to be discovered), and major advances in chemical engineering. However, there are limits to the economies of scale that can be achieved by increasing the size of plants, and in the 1970s the emphasis shifted to investment in improving the efficiency of existing plants through energy conservation and other measures. Third, substitution of plastics for other products was largely complete; indeed, some traditional products, such as in the paper and board industry, fought back with some success.

It is true that some emphasis was placed on what could be achieved in substitution terms by engineering plastics. However, the total world business was only worth $4200 million in the mid-1980s with about seventy producers to fight for it, giving about $60 million dollars of sales each; hardly a life-saver in chemicals industry terms (figures from Chem Systems International, quoted by Greenwell Montagu Research). Looking on the brighter side, although there are likely to be relatively few opportunities in future for the substitution of plastics for other materials (although PET continues to make progress as an alternative to glass bottles), there is likely to be little substitution of other materials for plastics. In other words, petrochemicals will not face the scale of decline in demand which the steel industry has suffered in the face of substitution from aluminium and plastics.

On the supply side the first factor that has to be taken into account is the failure of the industry to recognize that the years of rapid expansion were over, and that there was a need to restrain investment plans. These indications were already present at the end of the 1960s, but it was not until the first oil shock that it was widely recognized that previous growth rates could not be sustained. However even this episode provided only a temporary check to expansion. As an OECD study has noted: 'the investment decided prior to the first oil crisis had produced surplus capacity by 1973/1974. Subsequently, though production collapsed in 1975 and took three years to recover its 1974 level, production capacity steadily grew. So, instead of being reduced, the 1974 surplus got worse' (OECD, 1985: 83). The second oil shock led to a 'second false dawn . . . when fears that feedstocks would become both short in supply, and therefore expensive, led to a rapid but temporary increase in petrochemicals demands' (European Parliament, 1984: 22).

However, the underlying situation could not be evaded for long, and in 1980 destocking led to a sharp decline in petrochemicals output, initiating the most serious crisis the industry had ever faced. It must be said that a considerable proportion of the blame

Table 10.3 *New/announced capacities, new producers/ developing nations, 1982 onwards (thousand tons)*

	HDPE	LDPE	PP	PS	PVC
Middle East	360	905	112	152	450
Far East	605	850	744	73	222
Latin America	100	340	140	—	375
Eastern bloc	1265	1105	1380	118	668
Africa	555	700	311	—	360
Other	40	175	85	—	160
Totals of which	2925	4075	2772	343	2235
definite	1990	2620	2095	275	1304
Western Europe[a]	2300	5100	2800	2200	5300

[a] 1987 capacities.

Source: Greenwell Montagu Research.

for this crisis must rest with the industry. The companies failed to heed warning signals and continued to operate as if they were still in the expansionist environment of the 1950s and 1960s. Part of the problem was undoubtedly that the leading executives of the industry had begun their careers in that environment, and that companies were insufficiently flexible to adjust to changed circumstances (see Pettigrew, 1985).

The supply-side situation was worsened by the emergence in the 1980s of new producers of petrochemicals. European attention focused on the Saudi Arabians, but in fact they were only one of a number of new producers in the process of starting up operations (see Table 10.3). It should be noted that these new producers did not make an immediate impact on world markets, although even small landings of Saudi production in Western Europe had a disproportionate psychological impact, particularly in terms of unsettling prices. Indeed, it was widely believed in the industry that more damage was likely to be inflicted on the industry by panic reactions by European producers than by the Saudis themselves, who were committed to 'orderly marketing' and were expected to account for only 3 per cent of world ethylene production.

Nevertheless, the long-run implications of the emergence of the new producers were extremely serious for Western Europe. It should be borne in mind that the West European petrochemicals industry is highly dependent on exports: it exports some 50 per cent of its output, compared with around 10 per cent in the case of the USA. It is therefore vulnerable to being driven out of the third country markets, particularly the so-called 'deep-sea' markets, by new producers. In a number of key product areas the new and announced capacities of new producers exceed existing West European capacity.

Even allowing for the fact that some of these projects may never be undertaken, the situation is a serious one for the West Europeans. The developing world accounted for only 13 per cent of petrochemicals production in the mid-1980s, but has over 80 per cent of the world's oil and gas reserves. By 1995 Greenwell Montagu estimate that the 'rest of the world' could account for 23 per cent of world production, with Western Europe falling back from a third to a quarter. As a leading chemicals industry analyst, Stuart Wamsley, has commented: 'The traditional producers have to make room in the market place for the new producers, for by the 1990s the industry will have been through a third wave of development in terms of ownership and geographic location' (*Greenwell Newsletter*, 76: 18). This does not mean that all is lost for the European industry: as Wamsley stresses, the European market is a large one and sales growth could reach 6–7 per cent a year. However, a great deal depends on how the European industry copes with its overcapacity problem. By the mid-1980s there was increasing optimism in the industry that this task would be tackled effectively, although the improving situation in the industry 'prompted a spate of new investment proposals (an additional 17 per cent on existing capacity) which if realised could produce overcapacity problems in the early 1990s' (Wamsley, 1987b: 6).

Mechanisms for resolving the overcapacity problem

Unilateral action by individual companies

In an industrial sector in which large firms have a considerable measure of autonomy, individual plant closures by companies making decisions on the basis of their own assessments of likely future performance would be the normal response to an overcapacity problem. Such unilateral closures have taken place; for example, the Esso plant at Cologne. However, although they have eased the situation, the crisis could not have been resolved simply by companies responding independently to market forces.

There are a number of reasons why a solution through the market mechanism alone is not feasible. First, the market in petrochemicals is highly imperfect. As a British firm of stockbrokers has commented, 'the European market is not an ''open'' market in that most of the capacity is ''integrated'' either through long-term contracts or through internal requirements' (Wood–Mackenzie & Co., 1985: 5). In addition, there are at least three types of exit barrier which may discourage a company from leaving petrochemicals production: oil company ownership, state ownership and commitment to a particular pattern of production.

Oil companies may be willing to cross-subsidize their chemicals operations because they are adjacent to refineries, and closing the chemical plant might adversely affect the economies of scale of the refinery operation. Indeed, residuals from cracking operations may be fed back to the oil refinery (see Molle and Wever, 1984: 4). However, oil companies are not going to be prepared to subsidize loss-making operations indefinitely, and their presence in the industry may influence the timing of restructuring, rather than whether it occurs at all. An oil company may be able to afford to take a longer time to plan and execute a strategic readjustment of its chemicals operations.

State-owned and influenced companies accounted for around 40 per cent of West European commodity plastics capacity in 1985, compared with 30 per cent in 1978, although the level of state ownership fell as a result of privatizations in the late 1980s. This increase is largely attributable to the reluctance of the state-owned companies to close plants. They were more likely to give weight to the employment consequences of closure, particularly when plants had been constructed in particular locations as part of a regional strategy, and they were also influenced by political calculations about the desirability of having a domestically controlled petrochemicals capability.

However, by 1985 there were indications that their behaviour was beginning to change, and that the state-owned and influenced companies were at the forefront of a new round of rationalization. The proximate cause seems to have been a major deterioration in their cash-flow position. In the past this would have led to an appeal to the responsible national government for more funds. By the mid-1980s, however, national governments were less inclined to attempt to solve industrial problems by the provision of generous subsidies, and were putting a new emphasis on the importance of restructuring, often (as in Italy) with the cooperation of private-sector firms. Other developments included the complete reorganization of the Spanish state-owned companies, and the projected closure of the major Portuguese state-owned petrochemical enterprises.

The third type of exit barrier occurs when a company is locked into a particular chain of production. For example, ICI lost considerable amounts of money in the 1980s on PVC, but it did not exit from the market altogether. ICI's problem was that it had accumulated a chain of resources for PVC production which, in a sense, it felt obliged to use. It had substantial rock salt mines in Cheshire; through electrolysis, salt can be used to make chlorine which is needed for the production of PVC. In the 1970s it had opened a new PVC plant at Wilhelmshafen in West Germany as part of its strategy to penetrate the European market. ICI's eventual solution (discussed more fully later) was to rationalize PVC production through a joint venture with another major producer, ENI of Italy.

The exit barriers can, then, be said to have affected the speed at which adjustment occurred, rather than its eventual attainment. However, there was another problem which was related to the managerial culture of the industry. Petrochemicals managers had been used to a climate of continual expansion. Admittedly, by the late 1980s a new generation of managers was coming to the fore who were much more accustomed to handling the problems of contraction. Even so, in interviews I conducted with major companies in 1985, a recognition of the general seriousness of the industry's position was often offset by a tendency to believe that the blend of management skills, technological know-how, and product mix of the particular company would enable it to win through. In other words, there was something of a reluctance to translate the unpleasant realities of the general situation into decision-making at the level of the individual company. Pessimism at the sector level was seen as compatible with optimism at the company level. One should not exaggerate this point; many companies took some very difficult capacity reduction decisions in the 1980s. However, if matters had been left entirely to the companies, and if there had been no mechanisms for collective appraisal of the problem, the outcome might have been less satisfactory.

Acts of God

The nature of the chemical industry means that there is a very special form of unilateral removal of capacity: its destruction by accidental explosion, most appropriately referred to as an 'Act of God'. As it so happened, three such Acts of God played a significant part in securing temporary capacity reductions in 1985. In January the smaller of two ROW (BASF/Shell crackers) was damaged by an explosion. In May part of the cracker at the Essochem plant in Stenungsund (Sweden) was damaged by the collapse of a wooden cooling tower. Most significant of all, Enichem's plant at Priolo in Sicily, one of the largest and most modern in Europe, was damaged by a fire (the cause of which was not immediately identifiable), leading to a series of explosions.

These three accidents removed 8.5 per cent of European capacity, admittedly not permanently. These accidents led to a 'sudden rise in ethylene selling prices in both the spot and contract markets' (*Greenwell Newsletter*, 70: 5). Prices at the beginning of the year had been causing considerable concern in the industry, but although ethylene prices went up from DM1450 a tonne in May to DM1550 a tonne in August, following the unprecedented series of accidents, they had fallen back to the May level by October.

Agreements between companies

The most important contribution to resolving the problem has not come from autonomous responses by companies to market signals, but through negotiations between companies leading to bilateral agreements involving a reduction of capacity. There have been a number of such agreements which can combine a variety of elements: plant closures, transfers of ownership, and the formation of joint ventures. One of the first deals, that between BP and ICI, illustrates what can be achieved through a deal involving plant closures and transfers of ownership. In discussing this case it should be pointed out that the closures carried out by both companies were not provided for in the agreements made between them, 'but were, in practice, an inevitable consequence of their application' (European Commission, 1984: 71).

The deal was based on each company specializing in its area of strength in commodity plastics: ICI in PVC, and BP in LDPE. ICI sold its modern British LDPE plant to BP, and BP sold its most modern British PVC plant to ICI. In separate but parallel transactions, BP closed down its remaining, less efficient PVC plants. In addition, ICI took an extra 30 per cent share in an ethylene cracker previously jointly owned with BP, allowing ICI to close down an older, less efficient ethylene plant. 'The final result was a specialization of production in the UK with limitations on output through plant closures. An important and active producer was thus eliminated in the UK for each of the two main products concerned (PVC and LDPE)' (European Commission, 1984: 72).

This deal strengthened ICI's position in PVC, but it continued to make substantial losses on the product. In July 1985 it agree with Enichem that the two companies would merge their PVC interests in a joint venture, European Vinyls (EVC). A combined reduction in capacity was made possible: given that the initial combined capacity of the two companies was about a quarter of the European total, it was possible to make significant inroads on European capacity. Both EVC and other companies made capacity cutbacks. By 1988, EVC was making profits of $223 million a year.

Bilateral deals between companies allow them to specialize in their areas of strength, although one risk is that they may be areas of technological rather than commercial strength. The end result may be companies or joint ventures which are dominant in a particular product market, but are even less likely to make further reductions in capacity because of the size of their commitment. It should also be noted that the continuing use of such deals depends on the European Commission maintaining its benevolent attitude towards their potentially anti-competitive character (an issue discussed further below). Nevertheless, the overcapacity problem would not have been brought under control to the extent that it has been without the extensive use of bilateral deals between companies leading to reductions of capacity. This process offers an interesting example of what can be achieved through negotiations between companies, with state and international bodies taking a background role, in contrast to more traditional, tripartite forms of concertation.

Cartelization

The prewar chemical industry was characterized by international cartelization between the leading producers, Du Pont, ICI and IG Farben. Cartelization is much more difficult in postwar circumstances because the number of players on the market has increased significantly. Any cartel would be vulnerable to an early defection by one of its members. In any case, American companies are extremely nervous of doing anything which might be interpreted as a breach of anti-trust legislation, whilst the Europeans have to be mindful of Community regulations and likely action by DG IV of the Commission.

Nevertheless, the habit of cartelization in the industry has not been completely extinguished. In April 1986 the European Commission imposed fines totalling 57.85 million ECU on fifteen petrochemical producers who, the Commission claimed, 'operated a market-sharing and price-fixing cartel in the polypropylene market from 1977 in contravention of the competition rules of the EEC treaty' (Commission press statement, 24 April 1986). All the companies involved subsequently appealed to the European Court.

The Commission has admitted that 'in practice the cartelisation of the market was incomplete and did not entirely exclude the operation of competitive forces' (*Financial Times*, 15 January 1987). What is apparent from the documents seized in raids on the offices of chemical firms is that there was an elaborate structure of meetings between at least some of the firms at which prices were discussed. Although the nature of the arrangements varied over the six-year period, the ultimate shape of the structure was:

1 An unofficial directorate known as 'the big four' (set up in 1982). This was made up of the four largest producers of polypropylene.
2 The 'big four' met the day before the 'bosses' (directors or senior managers).
3 'Experts' (marketing managers) met twice a month in order to set 'target prices'.
4 'The producers also met frequently in local meetings in which they discussed the detailed implementation of the agreed measures in national markets, often down to the price of the individual truck-loads to each consumer' (Commission press statement, 24 April 1986).

In January 1987 the Commission carried out dawn raids on eight chemical companies looking for evidence of possible price fixing in polyethylene and PVC. Hoechst refused to recognize the Commission's search warrant and obtained an injunction against the search measures from the Administrative Court in Frankfurt on the grounds that the EEC inspectors should have a judicial warrant. The European Commission subsequently started a legal action against the government of West Germany for allegedly failing to enforce EEC anti-cartel rules, and imposed immediate fines of 1000 ECU a day on Hoechst. Hoechst mounted an action against the Commission, although in March it failed to obtain an injunction from the European Court to suspend the fines, or prevent the Commission from launching a search.

By artificially supporting price levels, cartelization discourages necessary capacity reductions. In any case, its effect has sometimes been undermined by price-cutting moves by companies which are intended to drive weaker competitors out of the business. One industry expert commented in interview, 'That never works. You don't drive people out, you drive prices down.'

Actions by national governments

Although the chemical industry is internationalized, actions by national governments can have an impact on the restructuring process, if only because of the importance of state holding companies in the industry. Nationalization of the French chemical industry initiated a process of reorganization, with Atochem emerging as the principal producer of commodity chemicals. There has been some restructuring of the French industry, and some industry commentators consider that the French have achieved more than they have been given credit for. An illustration is the merger of BP's polypropylene interests with those of Atochem, with Atochem holding a 51 per cent stake in the new company (Appryl). BP is to build a new plant near Marseilles, making use of Atochem's new catalyst technology, and BP will then close its existing plant in the area.

Where there are state-owned companies, national governments are more likely to have some involvement in the affairs of the chemical industry, but even non-interventionist governments may take decisions which have implications for the industry's structure. (See Chapter 5 in this volume.) The Thatcher government's passage of legislation creating an alternative basis for the valuation of ethane used for petrochemical purposes may be regarded as a technical change in tax law, but it led ICI to take a legal action against the government which was largely successful. Opponents of the move argued that an attempt to safeguard the viability of the new Esso–Shell plant at Mossmorran in Scotland placed ICI in a disadvantageous position. The government was clear in its own mind that it was remedying a tax anomaly rather than creating a state aid or subsidy. The merits of the argument about the case are not the concern of this analysis, but they do show that any national government may take actions which could have implications for industry structure and the resolution of the overcapacity problem.

The role of the European Community

It must be stressed that the European Community has not sought to take the kind of interventionist role it has performed in relation to the steel industry in the case of the chemical industry. The possibility of a 'crisis cartel' for the industry was raised in 1982, and there were discussions on the issue between the Commission and leading companies. There was some support for the idea from French and Belgian companies but, in general, it was rejected by the industry, and in particular by the influential British and German companies. Nor was the Commission very enthusiastic about becoming involved in sorting out the chemical industry's overcapacity problem. In part this was because they thought that the industry had created the problem for itself and should find its own solutions. In part it was because the employment consequences of substantial chemical plant closures were seen as of minor importance compared with the steel industry or, indeed, the synthetic fibres sector.

Nevertheless, the actions of the Community have had, and continue to have, a significant impact on the resolution of the overcapacity problem in two important respects. First, there is the question of the Commission's willingness to tolerate restructuring actions which could be interpreted as a breach of the Community's competition policy. Second, there is the question of the treatment of petrochemical imports from the Gulf countries into Europe.

In practice the Commission has taken 'a flexible approach towards joint structural capacity reductions aimed at achieving a healthier structural situation in the sector concerned' (European Commission, 1983: 55). The Commission recognized in the 12th Report on Competition Policy that a free market solution does not necessarily work in a crisis-struck sector. The least profitable surplus capacity may not be eliminated because of cross-subsidization within a company, or external subsidies. The Commission is therefore prepared 'to condone agreements in restraint of competition which relate to a sector as a whole, provided they are aimed solely at achieving a coordinated reduction of overcapacity and do not otherwise restrict free decision-making by the firms involved' (European Commission, 1982: 43). As an alternative to such sectoral arrangements, and one which has been the usual route in petrochemicals, the Commission has permitted 'agreements between a small number of firms providing for reciprocal specialization which would enable them to close excess capacity' (European Commission, 1982: 44). The Commission is, however, concerned to ensure that agreements of this kind do not lead to the elimination of competition, and that users would not be deprived of a workable choice of supplier. The process of giving approval for particular agreements between companies can be a long-drawn-out one, and will usually involve the European public affairs officers of the company concerned in efforts to ensure that the Commission is aware of the soundness of the their case.

As far as the question of imports from Gulf countries is concerned, it should be pointed out that the relatively limited quantities involved are not likely to have much impact on the European overcapacity problem as a whole in the medium term. Although countries such as Saudi Arabia clearly regard Western Europe as a strategic market, much of the impact of the Gulf countries will be in terms of displacing European exports from third country markets.

The issue has been a politically important one, both for the European petrochemicals industry and for the Saudi Arabians. The European reaction is in part explained by the disproportionate effect that relatively small landings of Saudi output have had on European petrochemical prices. The impact of the Saudis has been magnified by the psychological reaction of the Europeans who have a tendency to resort to rounds of price-cutting which are not justified by the circumstances. Indeed, a number of industry managers and experts suggested in interviews that the European industry had a tendency to search for scapegoats for its own problems, the Saudis fulfilling a role earlier performed by the East Europeans and the Americans.

Petrochemical products from the Gulf have very limited duty-free entry to the EC market under the provisions of the Generalized System of Preferences (GSP). This issue overshadowed the first ministerial-level talks between the EC and the Gulf Cooperation Council in October 1985. The Commission was aware that the Saudis might impose retaliatory tariffs on EC exports; the Saudis also played a clever card by emphasizing the interest of plastic processors in having cheap raw materials, pointing out that the processors employed far more people than the petrochemicals producers. In December 1985 the Community increased the ceilings below which duty-free imports were allowed, although not by substantial amounts: for the most competitive product, LLDPE, the increase was only from 6.1 million ECU to 6.6 million ECU. The LLDPE ceiling was soon exceeded, and the tariff reimposed on 6 January 1986; tariffs on other products were reimposed within a week. This led the managing director of the Saudi Basic Industries Corporation to complain of what he saw as a new hard-line stance by the EC, commenting that 'These new developments clearly indicate that the continuous protectionist sentiments and policies of the EC against Saudi Arabia's growing petrochemical industry still prevail, even though the kingdom continues with its liberal, free trade attitude towards EC exports' (*Financial Times*, 21 January 1986).

The petrochemicals industry argues that it does not advocate protection, but insists that the European industry takes only its fair share of Middle East output, and that it should not be 'disadvantaged by other major trading blocs adopting trading positions incompatible with GATT which mean that an excessive amount of Middle East production is diverted towards Europe' (CEFIC press statement, 28 November 1984). Unfortunately, there is always a risk that the particular interests of the industry may be sacrificed in the interest of maintaining harmonious relations between the EC and the Gulf countries. However, the cost advantage of the Gulf countries may have been exaggerated, particularly in the light of falls in the price of oil in 1985–6 which, in turn, brought down the cost of naphtha. The feedstock-based variable cost advantage of Saudi products is eroded when crude oil falls to $15 a barrel or below. Price rises following the Iraqi invasion of Kuwait in 1990 illustrated the continuing value of access to cheap feedstocks, although this was offset in this case by worries about the vulnerability of Saudi chemical plants to air attack. By 1990, the EC had agreed a common negotiating position in talks with the Gulf Cooperation Council on a free trade accord. A transition period of up to twelve years was envisaged for petrochemicals.

Facilitative associative action

The role of business interest associations in the resolution of the overcapacity problem has been left until last because of its importance in terms of the broader concerns of this book. There is no doubt that the players in the market do see a role for associative action. A director of ICI's petrochemicals and plastics division noted in 1982, 'the industry, through its associations, will have to become much more efficient and streamlined in dealing with perturbations from outside western Europe' (Beynon, 1982: 48).

The most significant development in the area of associative action has been the formation in 1985 of the Association of Petrochemicals Producers in Europe (APPE). APPE, which is a sector group of CEFIC, directly organizes the thirty-four largest petrochemicals producers in Europe. The companies concerned have demonstrated their commitment to APPE by appointing as their representatives the most senior executives in their organizations responsible for their petrochemicals operations.

APPE had as its predecessor an informal group of nine petrochemicals companies which funded a study of the overcapacity problem in the industry by two retired executives (the Gatti–Grenier team). This effort increased the availability of reliable and validated information about the extent of the problem, but CEFIC as an organization was ill-fitted to take up issues related to the overcapacity problem. As a federation of national federations, there was an underlying tension between the officials of the national organizations and the representatives of individual companies. Although it used to operate a Petrochemical Committee structure, its role was advisory or consultative rather than representational. As one of the reforms he sponsored during his presidency of CEFIC, Sir John Harvey-Jones of ICI encouraged the formation of APPE.

APPE has a number of objectives, including 'to study and to attempt to solve problems connected with the petrochemical industry'. Its initial tasks included the establishment of an industry statistical database and the assessment of the impact of petrochemicals from the new producers. It cannot, of course, itself restructure the industry; individual companies want to make their own commercial decisions, and the Commission would not encourage anything that resembled cartelization. What it can do is improve understanding of the underlying problems, develop an industry database, and provide a mechanism for representing the industry's views to the relevant authorities.

It could, of course, be argued that anyone with any connection with the chemical industry would be well aware of the extent of its overcapacity problem by the time APPE was formed in 1985. However, in the past the industry has been slow to react to evidence of overcapacity and, in particular, to treat cyclical upturns as a sign of a fundamental recovery. Thus, the role of APPE was even more crucial in the more optimistic conditions of the late 1980s than if the situation had remained as bad as it was a few years earlier. APPE will help to keep the overcapacity problem in the forefront of the minds of leading decision-makers in the industry. Much therefore depends on the quality of industry participants in APPE's work. In this connection it should be noted that its officers and steering committee were drawn from the following major companies in 1985: Atochem, BASF, BP Chemicals, Dow, DSM, Empetrol (Spain), Hoechst, ICI, Monsanto, Montedison and Neste Oy (Finland).

APPE's role should, then, essentially be seen as a facilitative one: focusing attention on the industry's problems, permitting informed discussion of possible solutions and, of course, representing the industry to the Commission. Its formation is an important step forward in the resolution of the overcapacity problem. However, it is important not to exaggerate what can be achieved through associative action. The main mechanism through which progress has been achieved is by bilateral agreements between companies, a process facilitated by the attitude of the Commission, and by the general awareness of the extent of the problem fostered by APPE.

Conclusions

In his introduction to this volume Martinelli draws a contrast between a company-led model of market-oriented adaptation and a 'European model' of greater state intervention and institutional density. In particular he raises the question of whether business interest associations are effective instruments for resolving such problems as excess capacity or whether firms, state agencies or supranational bodies are likely to accomplish these tasks better.

For reasons which have been outlined in this chapter, a simple model of autonomous adjustment in response to market signals would not work in the European chemical industry. If the only capacity reductions had been those undertaken on a unilateral basis, there would still be serious overcapacity in the European industry. The bilateral deals cannot be seen as simple market adjustments as they rely on often complex and long-drawn-out negotiations between companies, on the consent of the European Commission, and on the capacity of companies to look beyond their immediate commercial situation to a broader and longer-term picture.

However, nor can it be said that there has been a classic 'European model' of greater state intervention. National governments, although involved in the chemical industry through state holding companies, have not intervened in the industry to the extent that they have in sectors such as steel or even the auto industry. At the European level the Commission eschewed the option of a 'crisis cartel' to restructure the industry. When the then industry commissioner, Viscount Davignon met 'the nine' group of petrochemical companies in July 1982, a reasonable interpretation of his objectives would be as follows: he wanted to tell the industry that he did not envisage an administered solution of its problems following the precedents of steel and synthetic fibres; he hoped the 'nine' would exercise the leadership required to cause the industry to work out its own solution; and he wished to indicate that the Commission recognized that there was a structural overcapacity problem and hence that DG IV would be able to bless those deals (subject to certain provisos) which were aimed at correcting the situation.

Davignon made it clear that 'It is up to the industry to work for a consensus and to tell us which way it wants to go' (Davignon, 1982: 941). The industry did respond by developing greater self-governing capacities, first through the informal mechanisms of the 'nine' and the 'club' that funded Gatti–Grenier, and ultimately through the important institutional innovation represented by the foundation of APPE.

However, it must be made clear that the overcapacity problem was not resolved by associative action. Davignon drew a clear distinction between a sectoral approach, involving an industry agreement, and a bilateral approach based on inter-company arrangements. It was the latter route which the industry followed. Associative action contributed to a clearer and more informed identification of the scope of the problem. APPE also provided a forum within which the general contours of the problem could be discussed, and through which the views of the industry as a whole could be represented to the EC. Its contribution was, therefore, important but not decisive. Nor can it be argued that there were unrealized potentialities for associative action in this case; the route that was followed was broadly acceptable to the EC and the industry.

By the standards of industrial restructuring in Europe in the 1970s and 1980s, what has happened in the petrochemicals sector is something of a success story. For example, by 1985 'the imbalance between ethylene nameplate supply and demand [had] been reduced considerably and only a comparatively small amount of further rationalisation is needed' (*Greenwell Newsletter* 76: 16). Even in the difficult area of PVC there was evidence of a supply/demand balance by 1987. In the longer run, 'the increasing availability of products from the new producers will limit the growth prospects for the European producers of bulk chemicals and most observers feel that some further rationalisation may be necessary before the process of adjustment is carried far enough to ensure reasonable stability and returns' (Wamsley, 1987b: 4). By 1987 the priority was 'a further reduction in the number of producers [rather] than the straight elimination of capacity. Hopefully this can take place in part by the continuing withdrawal of the weak players as well as by far-sighted measures such as the ICI/ENI joint venture in PVC' (Wamsley, 1987b: 5). One problem is that the increasing age of petrochemical plants in Western Europe makes a scrap-and-build programme increasingly desirable: hopefully, the improvement in margins noted in 1987 will continue, and make such investments feasible.

There are, therefore, no grounds for complacency. Nevertheless, there can be some satisfaction about what has been achieved to date. Of course, the industry has not faced the substitution problems faced by industries such as steel, whilst closing a chemical plant does not have the employment consequences of closing a steel works. Hence, the external political pressures on the chemical industry are less intense, leaving it greater scope to work out and implement its own solutions.

However, even if the lessons to be learnt from the restructuring experience of the chemical industry cannot readily be applied to other industries, there are some observations which can be made in the context of the themes pursued in this book. What has happened in the chemical industry is neither an 'Anglo-American' nor a 'European' solution, but rather a distinctive blend of the two approaches which could properly be termed a 'chemical industry' solution. The chemical industry has a tradition of international cooperation between companies. International cartelization is no longer a viable option, but the links between companies, and in particular the relative denseness of its *international* associative networks, can be drawn on to support and facilitate negotiated inter-company solutions.

Adjustment therefore takes place between companies, but it does so in a context informed by associative activity, and facilitated by the actions of the European Commission. Associational development contributes to a pattern of negotiation which is company led and is blessed by the EC. It is difficult to think of a pattern of adjustment which would be more acceptable to the international companies which dominate the petrochemicals industry. These companies recognize the importance of associative action, as is shown by their active support for APPE. This is a case in which organized interests do have a role to play at the supranational level in relation to an industry adjustment problem. However, the role of the associational system is essentially a facilitative one, acting as a channel of communication between companies and the EC. It is the companies which make the crucial decisions on investment and disinvestment, but well-organized associations can contribute to the appreciative system of companies so that their decisions are made on the basis of a full understanding of all the dimensions of the overcapacity problem.

The limits of the lessons learned

In 1988 and 1989 the West European chemicals industry experienced a boom. Demand rose significantly, prices increased sharply, and profits improved markedly. By early 1988, West European production had reached nameplate capacity, and actually exceeded it in the winter of 1988–9. Work on the first new ethylene plant to be built in Europe for a decade was started at Antwerp by Neste of Finland and Petrofina of Belgium. Other companies announced significant expansions of existing plants, for example BP at Grangemouth. In addition, many existing plants experienced incremental improvements in capacity through debottlenecking, a practice referred to as 'capacity creep'. These announcements led to forecasts of ethylene capacity of 18.5 to 20 millions tons a year in Western Europe by 1995 compared with a nameplate capacity of 14.8 million tons a year in 1988 (*European Chemicals News Northern Europe Supplement*, July 1989: 4).

Outside Western Europe, the focus of attention switched from the Middle East to the Far East with substantial expansions under way in South Korea, Thailand, Taiwan, Malaysia and Indonesia. According to some forecasts, 40 per cent of the growth in world chemical production in the mid-1990s will come from Pacific Rim countries (*Financial Times*, 20 September 1990). With six new companies opening petrochemical plants, South Korean production of ethylene is expected to increase almost threefold to over 4 million tons a year by 1993.

Commodity petrochemicals production is a highly cyclical industry, and by the spring of 1990 there was evidence that a downturn was under way, prompted in part by slower growth in the world economy. All the major manufacturers reported lower profits in the first half of 1990, and there were increasing signs of overcapacity. By late summer, the position was complicated by the Gulf crisis producing sharp rises in the prices of naphtha, the principal feedstock for ethylene in Western Europe. The prices increased from less than $160 a tonne in July to $300 a tonne in September. However, a repetition of the crisis of the early 1980s was not anticipated. All the major firms have diversified from bulk chemicals to research intensive specialty products, which are less influenced by economic cycles and changing raw material prices. The real losers could be the ill equipped East European chemical plants who have already faced an increase in raw material prices following the Soviet decision to charge world market prices for oil from January 1991.

Commodity chemicals remain important to major companies as a source of cash flow and, in good times, substantial profits. The experience of the late 1980s shows that they may still be tempted into decisions that increase capacity, but the experience of the early 1980s has also made them more willing to reverse those decisions rapidly when circumstances change. In October 1990, Exxon Chemical announced that it was halting work on an expansion of the ethylene plant at Mossmorran while ICI announced that it was starting a two year 'belt tightening' period with a £100 million cut in its investment plans. The expansionist philosophy of the 1960s and 1970s will never return, and companies have a much more balanced portfolio of activities. Lessons have been learned, largely within companies themselves, but the industry's capacity to reflect collectively about its problems has also been permanently improved.

References

Beynon, D. (1982) 'Rationalisation: an ICI view', *Chemistry and Industry*, 16 January 1982, pp. 46–9.

Davignon, E. (1982) 'The position of the European Commission', *Chemistry and Industry*, 4 December 1982, pp. 938–41.

European Commission (1982) *12th Report on Competition Policy*. Brussels: Commission of the European Communities.

European Commission (1983) *13th Report on Competition Policy*. Brussels: Commission of the European Communities.

European Commission (1984) *14th Report on Competition Policy*. Brussels: Commission of the European Communities.

European Parliament (1984) 'Report drawn up on behalf of the Committee on

Economic and Monetary Affairs on the European petrochemical industry' (rapporteur: Mr Peter Beazley), Document 1-1108/83, passed in plenary session, 17 February.

House of Lords (1985) *Report from the Select Committee on Overseas Trade*, Volume II, *Oral Evidence*. London: HMSO.

Kenis, P. (1985) 'Industrial restructuring: the case of the chemical fibre industry in Europe'. EUI Working Paper 85/91. San Domenico di Fiesole: European University Institute.

Mény, Y. and Wright, V. (1986) 'State and steel in Western Europe', in Y. Mény and V. Wright (eds), *The Politics of Steel: Western Europe and the Steel Industry in the Crisis Years*. de Gruyter. pp. 1–110.

Molle, W. and Wever, E. (1984) *Oil Refineries and Petrochemical Industries in Western Europe*. Aldershot: Gower.

OECD (1985) *Petrochemical Industry*. Paris: Organization for Economic Cooperation and Development.

Pettigrew, A. (1985) *The Awakening Giant: Continuity and Change in ICI*. Basil Blackwell.

Wamsley, S. (1987a) *The Chemical Industry: The Current Situation*. London: Greenwell Montagu Research.

Wamsley, S. (1987b) 'The outlook for the chemical industry'. Paper presented at the Chemical Industries Association Business Outlook Conference, London, 15/16 January.

Wood–Mackenzie & Co. (1985) *The Impact of Saudi Chemicals*. Edinburgh, January.

11

Conclusion

Alberto Martinelli and Wyn Grant

This volume has dealt with the role of business interests associations in the chemical sector and has covered a variety of research questions which have been specified in the introductory chapter. In the conclusion we intend to focus on four themes which have emerged from the various chapters. First, the relative importance of associations versus other regulatory mechanisms in coping with chemical industry problems such as overcapacity and environmental protection, and in the relationship of business with government. Second, the relative importance of national styles of policy-making versus the features of the industry and the properties of policy areas in producing the conditions which are conducive to specific solutions to industrial problems. Third, the assessment of neo-corporatist theory in general, and of the Schmitter–Streeck model in particular, in the light of our research findings. Fourth, the effectiveness of the institutional density, concertation model versus the market-oriented, company-led model in coping with the problems faced by the chemical industry in the 1970s and 1980s.

In order to explore these four themes we will summarize some of the major findings of the volume. The first major finding is that associations are not the principal instruments of industry regulation and of business relations with government. The strategies of large firms which include direct relations with government agencies, friendly relations with unions, and agreements among firms which enhance the self-governing capacity of the industry, can be at least as important as, and often are actually more important than, associative action. This major role of large firms is related to such features of industry structure and market opportunities as size, pervasiveness and economic centrality of chemical firms both in national economic systems and in the world economy, and the high degree of capital concentration in the chemical sector, together with the internationalization of chemical trade and competition. In this respect this volume helps to modify the research framework developed by Schmitter and Streeck in the direction of emphasizing the role of the firm as an individual political actor.

This does not mean that business associations in the chemical sector are not important, or that they lack the organized complexity and the relative autonomy which are the necessary prerequisites for playing the role of a private interest government (see Chapter 7, by Jacek, in this volume). It means, rather, that here more than in other economic sectors associations often represent a second-best solution in dealing with the many problems the industry has to face. The specific nature of the problem area makes a difference: industrial relations are dealt with in a different way from overcapacity or environmental issues, but, in general, one can argue that associational development at the national and international level is not the chief instrument in solving sectoral problems. Rather, it plays a facilitating role in setting the pattern of negotiation which is company-led and under the benevolent eye of national governments or supranational institutions such as the European Community.

The way an industrial sector (as well as a national economy) is regulated, and the way it copes with its specific problems, are the result of a particular mix of a variety of types of coordination: market, state direction and regulation; managerial hierarchies; class-based, regional and policy communities; and business associations. These types of coordination are the institutional expression of the three basic principles of criteria of social regulation which can be identified in the theoretical tradition of the social sciences, in scholars as different as Thomas Hobbes, Adam Smith, Emile Durkheim and Karl Polanyi: the principle of authority or hierarchical control; the principle of exchange among at least formally equal partners; and the principle of solidaristic cooperation based on shared values and norms.

The different principles tend to identify more closely with one or the other institutional type of regulation: the principle of authority with the state and other bureaucratic organizations; the principle of exchange with the market; and the principle of cooperation with communities, clans and the like. But we consider it important to correct and supersede the typology in two ways. First, by distinguishing the state from managerial hierarchies which are usually conflated in one single type (see, for example, Hollingsworth and Lindberg, 1985). Second, by considering associations as a distinctive new type, as Streeck and Schmitter, with some uncertainty, tend to do (Streeck and Schmitter, 1984). In fact, although the infra-organizational and inter-organizational concertation which is characteristic of associations is the result of a special mix of the three criteria of authority, exchange and cooperation, associations provide a distinctive institutional mechanism of economic and social regulation and coordination. Organized interest associations are thus just one institutional mechanism of regulation and coordination of economic activities and industrial problems, alongside the market, the state, the hierarchical organization of the firms, and various forms of communities and clans.

There exists a widespread consensus among scholars who study economic and business questions that the self-regulating market, especially as it is conceived of in the model of perfect competition, has gradually lost its central and prevalent function in the coordination and regulation of economic activities, in comparison to other mechanisms, in particular the state and organizational hierarchies (see, among others, Williamson, 1975; Chandler, 1977). The state replaces the external impersonal mechanisms of the invisible hand with conscious regulation by political intervention aimed at stabilizing the economic cycle and at coping with the negative implications of the market for the social fabric, such as unemployment, relative income inequalities, pollution and the like. Organizational hierarchies internalize the rules of the market within oligopolistic business firms, reducing the degree of uncertainty, increasing the degree of predictability, and hence assuring the continuous flow of resources to the firms. Moreover, through various forms of joint or concerted action, they cope with problems of overcapacity, control of prices, allocation of raw materials, control of tendering, and such-like. Competition in the market continues to take place, but in the presence of active intervention by the state and of coordinated action by large business organizations.

A fourth type of coordination and regulation is that of communities or clans (Hollingsworth and Lindberg, 1985; Ouchi, 1977). This type refers to more traditional patterns of social relations which have been adapted to the needs of modern business. Communities or clans are informal organizations whose members, belonging to different formal organizations, develop long-term reciprocal expectations of exchange of economic advantages, power, influence and prestige. In certain circumstances they can both bypass complex market decisions and circumvent bureaucratic rigidities by relying on personal trust rather than the price mechanism or formal procedures.

Business associations are thus just one of the regulatory and coordinating mechanisms of an industrial sector; they are just one of the components of a regulatory mix. The importance of their role depends on a variety of variables related to the economic and political structure of a given country, the specific features of a given society, and the specific properties of a given policy area.

Most important among the sectoral variables are: the degree of concentration in an industry; the extent of internationalization of the relevant markets and of the structure of individual firms; the extent of public ownership; and how far the industry is fragmented into distinctive subsectors. Most important among the variables related to the nature of the political system and policy-making process are: the number of actors involved in the policy process; the relationship between the executive, the legislature and the judiciary; the degree of independence and importance of administrative agencies; the role of trade unions and key features of the system of industrial relations; and national patterns of policy implementation. Finally, consideration must also be given to variables related to the properties of different policies, such as the importance of a given issue for the political agenda; the extent of social concern and social visibility of a given issue; the number of stakeholders involved; the degree of technical complexity; and the availability of expertise outside the industrial sector and the business community.

We will briefly review these various variables with regard to the chemical sector in order to argue our second major finding, that is the fact that in the chemical industry sectoral variables tend to become more important than country variables in most policy areas. In other words, the strategies adopted by chemical firms and chemical business associations of different countries in coping with their problems and in dealing with their relevant partners, especially the state, tend to become more similar. However, there are important exceptions to this general pattern.

Perhaps the most persistent case of national variations outweighing sectoral variations is that of industrial relations. As Whitston points out in Chapter 8, the legal and institutional structures of industrial relations are chiefly shaped by national rather than sectoral concerns. Thus, in contrast to West Germany, union involvement in public policy-making in Britain has been deeply influenced by the 'voluntarist' tradition of collective bargaining. Other variables which lead to cross-national differences in the industry will be discussed more fully below.

Distinctive sectoral variables

What are the principal variables which give the sector its distinctive character? The industrial chemical industry has a long history of oligopolistic patterns of competition, permitting, and reinforced by, a propensity to international cartelization. The highly integrated production process of the industry requires a considerable start-up investment, providing formidable entry barriers which can be surmounted only by firms which have accumulated capital elsewhere (as in the case of the oil firms which entered the industry in the postwar period), or by a state backed firm (as in the case of ENI in Italy). Another distinctive feature of the industry is its high level of internationalization, a long-standing feature which has been accentuated in the 1980s by the increasing penetration of European chemical companies in the US market through the acquisition of American companies. As Mussati and Soru observe in Chapter 2, one of the factors leading to the internationalization of the industry has been its research intensity. Given the cost of research and development activities, and the market lead which can be provided by a breakthrough, there is an incentive to exploit new products and processes on an international scale.

The international character of the industry gives rise to, and is sustained by, an international management culture, with many opportunities for managers from different firms to meet at a variety of international conferences, ranging from the formal and public, to the informal and secret. The traditional management culture of the industry has been technologically based and production-oriented, with the discipline of chemical engineering playing a key role (see, for example, the discussion of the switch to large, single-train plants in Spitz, 1988: Ch. 11). However, this orientation has been changing under the greater competitive pressures of the 1980s, and the increasing importance of specialty chemicals which have to be tailored to customers' needs, as distinct from commodity products. Individuals with a marketing background have assumed a greater role in a number of companies. Even so, chemical firms tend to continue to specialize in the chemical industry, and not to stray outside it (leaving aside the involvement of oil companies), even in countries such as Britain and the US where conglomerates have become an increasingly important form of corporate organization.

Because of the importance of specialized technical information in the industry, national bureaucracies staffed by generalist civil servants have generally had a limited capability to comprehend developments in the industry. They have generally been reliant on firms and associations as sources of information. Something of an exception to this pattern is to be found in international agencies such as the OECD, EC and specialized UN agencies concerned with chemical transportation issues. Even here, personnel, information and expertise resources are limited compared to those of the industry.

However, it has become less possible to handle the industry's relations with government through routinized relations with non-expert bureaucrats as the industry has acquired a higher public profile in all countries as a result of increasing public concern about environmental issues. In all countries the industry faces a political challenge in the environmental area. Moreover, incidents in one country have an immediate impact on public opinion in other countries, affecting the industry worldwide. In dealing with the environmentalist challenge, the industry is helped by having relatively accommodationist unions which are prepared to make common cause with the employers against environmental movements, as can be seen most notably in West Germany. More generally, the accommodationist character of the union movement means that management has not generally faced the industrial relations problems that have afflicted other sectors such as the auto industry.

Differences in national political systems do, nevertheless, have a significant impact on the operating environment of the industry, and it is to an examination of those variables that we now turn. We start with what is possibly the most significant differentiating variable: the nature of the party system.

National party systems and the chemical industry

We are concerned with two aspects of the party system as differentiating phenomena. First, the composition of the party system in terms of the role and influence of parties that take a strong stand on issues which are of central importance to the chemical industry such as environmental policy. Second, the distinctive features in different countries of the relationship between the party system, government and the economy. The two aspects are interrelated: if a particular system has within it parties that are particularly interested in environmental issues, and if that system is characterized by 'party government', then the impact of political parties on the industry is potentially considerable. However, account also has to be taken of the effectiveness of government. In Italy, public interest in environmental issues is increasing, political parties are very strong, but the delivery capacity of government is limited. Hence the chemical industry, although facing a more difficult political environment, still retains a greater autonomy of action than in countries where implementation mechanisms are more effective.

An analysis of the party system must take account of the dynamics of electoral competition which lead existing parties to modify their policies, and to seek to impose stricter controls on the chemical industry, in order to counteract the appeal of the new environmentalist parties. One must also take account of the fact that such parties may be represented in regional or local governments (for example, for a time the Greens held the environmental portfolio in the Land of Hesse, and in 1987 held two seats in the city government of Milan).

The extent to which party systems differ in their relationship with the chemical industry can be observed through a comparison of Britain, Italy and Germany. The British case is characterized by the relative lack of engagement between the industry and the party system. Industrial chemical companies (as distinct from pharmaceutical companies) are not major donors to the Conservative Party; ICI in particular has a policy of not making political donations. In a first-past-the-post electoral system the influence of the Greens is limited. All parties have a vague commitment to environmental improvement, but the practical effects in terms of constraints on the chemical industry are somewhat limited. The main pressure for change comes from EC legislation (see Chapter 9).

In Italy, parties tend to control sectors of civil society, and colonize state institutions. Especially in the late 1960s and 1970s the chemical industry became an arena for inter- and intra-party competition. This had a negative impact as it delayed an overdue restructuring of the Italian chemical industry. By the late 1970s the influence of parties as parties was diminishing, and government was able to take a more objective and detached stance towards the problems of the industry (see Chapter 4).

For the entire period of the Federal Republic, with one exception, West Germany has had coalition governments. This creates a need for greater engagement with the party system than in Britain with its single-party governments, but still less than in Italy with its multi-party system with much lower entry barriers than in West Germany. In the 1980s in West Germany the Greens have had a major impact on the party system in terms of encouraging policy adjustments by the existing parties. For example, the SPD started to formulate a chemical industry policy which was, in fact, a policy against the chemical industry. This policy was moderated by IG Chemie working within the SPD (see Chapter 9).

Given the growing public concern about environmental issues we can envisage a growing importance for parties as actors in relation to the impact of such issues on the chemical industry. However, different electoral systems, and other factors, mean that there are considerable variations from one country to another in the strength of such parties. Hence the party system remains a differentiating variable.

Industrial policy

In the postwar period, industrial policy in the chemical industry varied from state ownership in Austria, France and Italy; through active regional policies in Britain and Italy; to creating conditions in which the industry could be successful, as in West Germany; and the virtual absence of an industrial policy in the United States (see Chapter 5, by Adams). In France and Italy the need to catch up with more established producers, and the tradition of government intervention, led to specific state plans for the expansion of the chemical industry, and also to an active role for the state as entrepreneur. This involved fostering national champions such as ENI, and also a close linkage between corporate strategies and government goals. The top management of such state-owned enterprises was appointed by the government. Such countries have used the whole range of industrial policy instruments from regional assistance to research and development incentives. As Adams points out, in France and Italy the state was often forced to intervene directly in the industry to salvage operations that had been created by past policies. At the other end of the spectrum, the United States has not had state-controlled firms or a regional policy, and has left the development of the industry largely to market forces. Up until the Tokyo round the main form of industrial policy was protection in the form of the American Selling Price (Ilgen, 1983: 676). Britain provides an intermediate case with a government-influenced national champion that was not state-owned (ICI), and effective regional policy used by the chemical industry, and limited funding for research and development.

It should be noted, however, that industrial policy may be less of a differentiating factor than in the past. In the 1980s, governments such as those of France and Italy have moved towards more market-oriented solutions to the problems of their chemical industries, a development consistent with the completion of the internal market and the generally liberal stance of the EC. As Adams points out, the emphasis is now on restructuring rather than development.

The legislative and administrative process

Another set of important factors relates to the legislative and administrative process. In a federal system, provincial governments can be important actors in relation to the chemical industry. They may have their own industrial policies; they are often responsible for the implementation of environmental regulations; and they may lobby for industries located in their regions at the national level. In the case of Canada, as discussed in Chapter 7, by Jacek, serious rivalries developed in the recessionary conditions of the early 1980s between the gas-based producers in Alberta and the oil-based manufacturers in Quebec. In the case of Germany, chemical firms attach considerable importance to the development of working relationships with Land governments. Even in more centralized systems, local governments can make significant decisions affecting the chemical industry, for example in relation to planning laws and environmental issues. In Italy, in November 1987, a Montedison plant's operations were prohibited by a local authority, and the firm closed down most of the plant. A sharp conflict ensued involving the plant workers, the 'greens', the local authority, the government, the unions and the employers' associations.

The importance of the legislature in making decisions affecting the chemical industry varies considerably from one country to another. In the United States, House and Senate committees are important both in defining issues relating to the industry through the hearings process, and in drafting legislation that sets parameters for the activities of the industry. At the other end of the spectrum, the British Parliament usually endorses legislation produced by the government; the main exception to this pattern is private members' legislation, relating to such subjects as the use of animals in laboratory experiments. Italy and Germany are intermediate cases. In Italy, most laws are shaped by Parliamentary committees. In the case of *decreto legge* (law by decree) which has increasingly been used in recent years, in order to counterbalance the slow pace of the legislative process, Parliament retains a veto power. Italy is thus closer to the American end of the spectrum; whereas Germany is closer to the British end. The German Parliament has had a somewhat larger role in relation to the chemical industry than that of Britain (see Grant et al., 1988).

Most countries have independent regulatory agencies that are concerned with environmental and/or health and safety questions. However, such agencies have a particular significance in the United States, with the Environmental Protection Agency and the Occupational Safety and Health Administration being particularly significant in relation to the chemical industry. As Brickman et al. note (1985: 77), 'The legislation conferring these extensive powers leaves the administrative agencies free to choose from a wide range of technically and economically feasible control strategies.' They conclude that 'In the United States . . . the social regulatory agencies of the 1970s can interpret law and develop policy with very little interference from executive departments with countervailing policy interests' (Brickman et al., 1985: 79). Agencies in other countries are more constrained.

The role of experts

In an industry such as the chemical industry, scientific experts are bound to play a significant role in the formulation and implementation of government policies affecting the industry, particularly regulatory policies. However, there are substantial variations in the role that experts play from one country to another. The most striking contrast is between the United States and Germany. In Germany there are close links between the industry and the scientific community. Companies have usually been headed by chemists, often holding university professorships. The industry has a fund supported by a levy which is used to assist basic scientific research in universities, and to award prizes and grants to promising doctoral students. In general, the scientific community has been supportive of the industry, and the legitimacy of scientific expertise has been used to defend it against environmental critics. The resources available to the independent regulatory agencies in the United States, the role played by public interest groups, the value attached to expertise in the society, and the tradition of subjecting expert testimony to detailed public scrutiny have created a very different situation there. There is less concentration of expertise in official advisory committees than in European countries with more use of outside consultants (Brickman et al., 1985: 168). They conclude:

> Since the United States makes the most extensive use of all kinds of expertise and provides the most thorough support for its regulatory actions through formal analytical argumentation, it is also in this country where the value and credibility of various kinds and means of analytical input are most heatedly debated. In Europe, the regulatory role of expertise rarely takes on the same political overtones. (Brickman et al., 1985: 180)

The judiciary

In the United States the judiciary has a crucial role in making decisions that affect the chemical industry. Brickman et al. note:

> American federal judges are prepared to look more critically at the substantive basis of regulatory decisions than are the administrative courts of France and Germany. . . . American courts also exert considerable indirect influence on the procedures of federal regulatory agencies. (Brickman et al., 1985: 110)

In general the courts played a more limited role in relation to the chemical industry in Western Europe and Canada than they do in the US. However, the European Court has been an arena for appeals by chemical companies against actions taken by the European Commission in the area of competition policy. The treaties have also provided a basis for appeals in national courts against decisions made by national governments. For example, ICI made an application for judicial review of section 134 and schedule 18 of the Finance Act 1982 on the grounds that the action contravened provisions of the EEC Treaty. Given the importance of Community decisions in relation to an internationalized industry such as chemicals, the importance of judicial review in Western Europe may increase, although not to the extent that has long characterized the United States.

The coordinating role of business associations

The extent to which political action is coordinated through business associations varies from one country to another, reflecting national variations in the way in which business seeks to achieve its political goals. Austria and West Germany have well-established traditions of associative action, and in the chemical industry in those countries the associations play the leading role, although the largest West German firms have recently appointed their own European coordinators. In the Anglo-Saxon countries (Britain, Canada and the US), the government relations divisions of individual firms are significant actors alongside the associations. In Italy, in terms of impacts on policies which are important for the industry, direct contacts are as important as associative action. ENI is a very special case because it has its own association representing the firms of the ENI group.

It must be emphasized, however, that the role of business associations differs from policy area to policy area across countries, as well as between countries. In all countries, associations play a significant role in attempting to influence the development and implementation of environmental policy. By way of contrast, as Chapter 10 by Grant shows, the firms themselves played a crucial role in coping with the overcapacity problem. Agreements between firms were of a commercial character and took place away from associations, although organizations such as APPE have played a role in improving understanding by firms and the EC of the scope of the overcapacity problem. Trade policy issues are a particular responsibility of the European-level organization, CEFIC, given the commercial policy powers of the Community. The involvement of associations in research and development issues is more nationally differentiated, with the German industry leading in terms of collective action with its fund for chemical research.

In spite of the intrusion of these differentiating characteristics across countries, linked to the character of their political systems, our general conclusion is that the chemical industry is one in which sectoral variables are of predominant importance. There is a remarkable homogeneity of problems, and of solutions to problems. Even where significant differences existed in the past, they are rapidly diminishing (see Chapter 4). The most important explanations of sectoral convergence are the degree of concentration in the industry, the high level of development of inter-firm relations, and growing political concern about environmental, health and safety questions which have an increased salience at least in advanced industrial countries. These tendencies are further reinforced by the increasing extent to which issues are handled by international institutions such as the EC.

Neocorporatist theory

We now turn to our third theme: what does the volume tell us about neocorporatist theory and the framework developed by Schmitter and Streeck for the international business interest project?

Neocorporatist theory has had a number of different formulations by scholars such as Schmitter, Lohmbruch, Winkler, Cawson and Crouch, among others. A useful synthetic definition is provided by Jordan and Richardson:

> The concept assumes that interest groups do not merely attempt to influence governmental actions, but themselves become part of the decision-making and implementation system. In return for this participation in policy-making the groups – through the control of their members – make society more manageable for the state or government. (1987: 18)

This is not the century of neocorporatism in chemicals. Basic elements of the neocorporatist model, such as intervention, intermediation and incorporation, are not well developed, and there are many features of the industry and its political circumstances which are distinct from the neocorporatist model. These include the role played by large firms in the policy-making process, including direct resolution of problems between firms (inter-organization concertation), the continuing importance of legislative bodies, and the role of regulatory agencies and the judiciary in the policy-making process (that is, the exercise of hierarchical authority instead of concertation). This does not mean that we think that, even for the chemical sector, a pluralist model would work better.

A third picture emerges, which could be called a 'chemical industry solution'. This is characterized by a complex network of negotiations which take place between a considerable number of actors and which take significantly different forms in the United States and Western Europe. As pointed out in Chapter 10, adjustment takes place between companies, but it does so in a context informed by associative activity and facilitated by the European Commission. Associational development contributes to a pattern of negotiations which is company-led, endorsed by national governments and blessed by the European Community.

What does the framework tell us about the more specific theoretical framework developed by Schmitter and Streeck, in which they drew on neocorporatist theory and organizational sociology? These questions have already been discussed in the introductory chapter. Two basic assumptions of the framework state that there is a correlation between organizational development and private interest government by associations, and that greater organizational development of associations indicates that associative action is more important to business than other means for defining and expressing its interests. Organizational development is defined in terms of the capability to organize complexity, and to obtain relative autonomy from the state:

> Organizational structures are the more 'developed' the more encompassing they are in scope and purpose (the more external effects and interdependencies they internalize); the more specialized and coordinated they are internally; the more safely their supply of strategic resources is institutionalized; and the greater their autonomous capacity to act and to pursue long-term strategies regardless of short-term environmental constraints and fluctuations. (Schmitter and Streeck, 1981: 124)

The studies contained in this volume fail to confirm these hypotheses. Business associations in the chemical industry form a rather coherent and well-organized associational system. Highly integrated production processes, geographical concentration of industries, and the dominant position of large firms contribute to greater homogeneity (there are fewer first-order associations than in other industries, only four countries out of nine have territorial associations, and only three out of nine have associations specialized by firm size). But in spite of that, chemical business associations are less important than autonomous action by large firms in expressing industry interests, and act as private interest governments only in a limited way. Such a possibility was, indeed, anticipated in the original research design when it was hypothesized that 'Where resources are very unequally distributed – say, between one large leading firm and many equally small ones – there will be a strong temptation for the dominant actor "to go it alone" in direct, unintermediated action' (Schmitter and Streeck, 1981: 64). Chapter 7 provides some examples of private interest government activity in the chemical industry, but they are more limited in extent and importance than the examples to be found in other sectors (see Jacek, 1987). This happens because, even when associations are potentially fit to play a private government role, firms may prefer direct action in the market, concertation among themselves, and direct dealings with governments. The existence of a core number of large firms, or the degree of economic concentration of an industry, is thus a crucial variable in explaining the actual scope and importance of associational action.

Market competition versus institutional concertation

Our fourth theme concerns the choice between the institutional density concertation model and the market-oriented, company-led model as a means of coping with the problems faced by contemporary Western economies. We are aware of the fact that the 1980s are considered by many scholars to be a period when the market has been rediscovered not only as a superior mechanism for the most efficient allocation of resources, but also as the best means to attain social goals (for an exposition of this view, see Brittan, 1988). We are also aware of the fact that when confronted with criticisms about the possible social costs of business activities some writers argue that all that is necessary is a general framework of laws and legislation.

Our research findings suggest a different view. An excessive reliance on market competition among individual firms would not have produced an orderly reduction of capacity in the West European petrochemical industry (see Chapter 10). Nor would a reliance on a free labour market or an effort to dramatically reduce the bargaining power of trade unions have produced benefits of long-term social peace enjoyed by the German and other chemical industries (see Chapter 8). Moreover, attempts at an adversarial form of environmental and safety regulation, as practised in the United States, are often costly in procedural terms and do not necessarily produce better results than agreements based on concertation between government and relevant interests (see Chapters 7 and 9). A final example is provided by cooperation between firms through a fund linked to their associations to support basic scientific research in the German chemical industry (see Chapter 7). One of the benefits of such an arrangement is that it develops a close working relationship between the firms and the scientific community, and that it increases the resources available to scientific sectors which are often financed only by governments and public bodies.

These examples and others which can be drawn from our volume lead us to three major conclusions which permit a reframing of some of the hypotheses outlined in the introduction. First, market adjustment and concertation are to be considered as ideal types which are intermeshed to a varying extent in reality. This is especially the case in the chemical industry because of the crucial role played by large firms. Second, in the 1980s, the two ideal types of market adjustment and concertation have tended to become closer. In countries such as Italy, institutional density has been maintained, but the decision-making environment of firms has become less politicized, and the pressures originating in international competition to adapt to changing market conditions have intensified. On the other side, new institutional mechanisms have been created at the European level, notably the Association of Petrochemicals Producers in Europe. Even in the US and the UK, where the market adjustment model seems dominant, certain chemical industry issues, notably the environmental issue, require negotiations among large numbers of actors. Third, the increasing complexity of the decision-making environment of companies, and the growing numbers of relevant stakeholders, lead to a situation in which the price mechanism cannot be relied upon as the only complexity-reducing mechanism. Rather, a changing mix of institutional forms of coordination of the economy, adaptable to specific circumstances and needs, is required.

The formation of Enimont, the largest Italian chemical company resulting from the merging of ENI and Montedison activities in important areas of the chemical industry, provides the most recent, although shortlived, example of this kind of institutional mix. It was the result of the combined operation of different institutional forces: increasing competition in the global markets for chemicals and corporate strategies for rationalizing production (see Chapters 2 and 4); the Italian government's preference for the formation of a 'national champion' in a strategic industrial sector; the long, direct negotiations between the two companies facilitated by changes in the control group of Montedison; the existence within the Italian government of a specific policy community, favourable to this arrangement; and, finally, the fact that Federchimica was the first Italian business association to affiliate both private and public firms.

In the contemporary chemical industry dramatic changes in the continental European model of institutional density and concertation do not seem necessary. Business interest associations remain an important component in the institutional mix which facilitates the operation of a negotiated economy. However, global firms responding to changing international market conditions and engaged in a complex network of relationships with other firms are the crucial actors.

References

Brickman, R., Jasanoff, S. and Ilgen, T. (1985) *Controlling Chemicals: The Politics of Regulation in Europe and the United States*. Ithaca: Cornell University Press.

Brittan, S. (1988) *A Restatement of Economic Liberalism*. London: Macmillan.

Chandler, A.D. Jnr (1977) *The Visible Hand: The Managerial Revolution in American Business*. Cambridge, MA: Belknap Press of Harvard University Press.

Grant, W., Paterson, W. and Whitston, C. (1988) *Government and the Chemical Industry: A Comparative Study of Britain and West Germany*. Oxford: Clarendon Press.

Hollingsworth, J.R. and Lindberg, L.N. (1985) 'The governance of the American economy: the role of markets, clans, hierarchies, and associative behaviour', in W. Streeck and P.C. Schmitter (eds), *Private Interest Government: Beyond Market and State*. London: Sage. pp. 221–54.

Ilgen, T. (1983) '"Better living through chemistry": the chemical industry in the world economy'. *International Organization*, 37: 647–80.

Jacek, H.J. (1987) 'Business interest associations as private interest governments', in W. Grant (ed.), *Business Interests, Organizational Development and Private Interest Government: An International Comparative Study of the Food Processing Industry*. Berlin: de Gruyter. pp. 34–62.

Jordan, A.G. and Richardson J.J. (1987) *British Politics and the Policy Process*. London: Allen & Unwin.

Ouchi, W.G. (1977) 'Review of *Markets and Hierarchies* by O.E. Williamson', *Administrative Science Quarterly*, 22: 540–4.

Schmitter, W. and Streeck, P. (1981) 'The organization of business interests: a research design to study the associative action of business in the advanced industrial societies of Western Europe'. (Berlin: International Institute of Management, labour market policy discussion paper).

Spitz, P.H. (1988) *Petrochemicals: The Rise of an Industry*. New York: John Wiley.

Streeck, W. and Schmitter, P. (1984) 'Community, market, state and associations? The prospective contribution of interest governance to social order'. European University Institute Working Paper No 94.

Williamson, O. E. (1975) *Markets and Hierarchies: Analysis and Antitrust Implications*. New York: Free Press.

Index

The abbreviation BIA denotes business interest associations.
References to tables appear in italics.